G000167960

Margins in European Integration

Margins in European Integration

Edited by

Noel Parker
Senior Lecturer in European Politics
University of Surrey, Guildford

and

Bill Armstrong
Research Scholar
University of Surrey, Guildford

First published in Great Britain 2000 by
MACMILLAN PRESS LTD
Houndmills, Basingstoke, Hampshire RG21 6XS and London
Companies and representatives throughout the world

A catalogue record for this book is available from the British Library.

ISBN 0–333–74710–0

First published in the United States of America 2000 by
ST. MARTIN'S PRESS, INC.,
Scholarly and Reference Division,
175 Fifth Avenue, New York, N.Y. 10010

ISBN 0–312–22958–5

Library of Congress Cataloging-in-Publication Data
Margins in European integration / edited by Noel Parker and Bill Armstrong.
p. cm.
Includes bibliographical references and index.
ISBN 0–312–22958–5
1. Europe—Economic integration. 2. Europe—Boundaries. 3. European federation. I. Parker, Noel, 1945– II. Armstrong, Bill, 1954–

HC241 .M2734 2000
337.1'4—dc21

99–059898

This book is printed on paper suitable for recycling and made from fully managed and sustained forest sources.

10 9 8 7 6 5 4 3 2 1
09 08 07 06 05 04 03 02 01 00

Printed and bound in Great Britain by
Antony Rowe Ltd, Chippenham, Wiltshire

Contents

List of Tables

List of Figures

Acknowledgements

The meetings behind these papers began in April 1997 by the European Borders Group of the Department of Linguistic and International Studies of the University of Surrey. The first meeting enjoyed the support of the University Association for Contemporary European Studies. Subsequent meetings have been funded with the generous support of the Department of Linguistic and International Studies. Dr Parker received additional help in developing the themes of the book and during the editing from the Dansk Udenrigspolitisk Institut in Copenhagen and the Center for Kulturforskning of the University of Århus. We are also grateful to Professor Clive Archer, Professor Clive Church and Dr Hans Mouritzen, who kindly read parts of the text in draft and offered much helpful advice on how to improve it.

N.P.
B.A.

Preface

The papers in this volume started life in a discussion held in April 1997, at a seminar under the title '"Ungovernable" Margins: Governance and Integration on the Edges of Europe'. We began with a problem that was, and still is, visibly present in the politics of integrated Europe. By 'integrated Europe' we had in mind not simply the territory formally incorporated into the European Union, but that embraced by a patchwork of integrated entities, such as the NATO alliance or the Schengen Agreement. This 'integrated' Europe – the EU, NATO, the Conference on Security and Cooperation in Europe (CSCE), the Council of Europe, the different national political orders and, for that matter, many sub-national ones too – is experiencing serious difficulty in organizing processes for the management of the common business, that is the governance, of society in the territory of Europe. They find this particularly difficulty in those territories where the tight, pre-existing framework disappeared in the early 1990s – though the problem was encountered in other parts of the continent in earlier years. So our agenda first concentrated attention on those areas in Central and Eastern Europe, the potentially 'ungovernable' margins of our title. We asked how viable were the attempts of the EU and others to organize integrated governance for these areas.

But we were also aware that this was not the only territory where integrated governance might be problematic. So we drew other margins into our ambit at an early stage. From a clear starting point – the difficulty that the Western European 'centre' has in extending integrated governance at its perceived margins – we moved rapidly backwards: that is to say, we found the widely accepted starting assumptions too questionable to stand. Why, for example, were some areas margins in the first place and others centres? We had initially framed the problem in terms of the will of the West to arrange things on *its* margins. But, while such rhetoric is easily found in the public statements of the European Council, etc., why were *we* assuming that the margins were a 'problem' which it was the business of the centre to 'solve'? Is it so obvious that order will be brought to the margins from the centre?

Discussion moved to a different perception: of an integrated Europe with many inveterately loose ends – margins in the sense of territories, people and processes that defy thorough inclusion in an effective Europe-wide order. In a longer historical frame, one might add, loose ends in political structures may have been an incurable feature of Europe, and even one of Europe's key geopolitical assets (Østergård, 1998). So, in this way, though in some senses going backwards, we were also changing the focus, so as to open up our initial question about governance at the margins to the *two-way* relationship between integrated Europe and what is thought of as its margins. Rather than estimating the prospects for integration's being extended to the margins of Europe, our discussions became an exploration of the possibilities for researching European margins and marginality as such.

This complicated considerably the import of the term 'margin' – in ways that Chapter 1 attempts to grasp more fully. On the one hand, the range comprehended by the expression broadened. There would be margins to consider not only on the territorial outside of 'Europe', but also on the *inside* of the integrated European order. Hence the carefully chosen preposition of our title: margins 'in' European integration, not margins 'of' European integration. Indeed, properly speaking, margins might have to be identified not with territory at all, but with socioeconomic *practices* that are awkward in relation to the socioeconomic order that is seeking to subsume them. The misfit is what most obviously makes the margins marginal.

Shifting attention to a two-way relationship between integrated Europe and its margins opened up a number of little-considered possibilities. We could expect to find some instances where, against the balance of forces suggested by the connotations of the term, the so-called margins exercised a degree of conscious power over the centre. We could also expect instances where the margins had an impact upon the integrated Europe regardless of any deliberate intentions, preferences or power on the part of actors on the margins. Finally, it could be that the very fact of the margins' marginality, by challenging or worrying the structures at the centre of the system of governance, produces effects in the system overall.

A margins research agenda

It is not new to assert that state boundaries, the lines that demarcate one state territory from another, are currently changing in character, if not simply losing their significance. In Europe, indeed, this is seen partly as the result of integration itself. For integration – like the regionalization which has been a contemporaneous geopolitical trend associated with, or even attributable to, integration – diminishes the meaningfulness of previous borders, undermining demarcations and overturning powers, notably governmental controls, which have been articulated at borders. Malcolm Anderson (1996) has accordingly painted a picture of liberal, Western states working hard to maintain the integrity inherited from their successful boundary-drawing in the past, while other political and cultural forces, acting in the minds of peoples, draw different boundaries according to other logics.

Thus it is that the declining significance of earlier borders falls as well within a group of phenomena which give many the impression that only processes and values deriving from the non-local, 'global' level can stand in the coming world. Yet, if European integration has been a purveyor of such effects, it has also been a victim. It seems inconceivable that the integrated Europe will ever replicate the political, cultural or economic standing and territorial integrity of the national state which globalization has put on the defensive. So boundary-making is likely to be a continuing preoccupation in European integration. And this suggests that marginality too may be a chronic condition, or a permanently significant feature of integrated Europe. There are thus, it emerges, three kinds of reason that lend promise to researches on integrated Europe pursued from the point of view of its margins.

First, there is a common-sense observation. Especially since the end of the Cold War it appears that, where integration happens, it does so on a number of different margins, all at the same time. There has been a proliferation of boundaries for Europe, deriving from the parallel extensions of different integrative bodies. The Council of Europe has been greatly enlarged. NATO has ingeniously created, in Partnership for Peace (PFP), a separate ante-chamber to itself to embrace ex-Soviet-bloc states *pro tem*, while the new situation evolves and the parent body makes up its mind what to do. The CSCE has attempted a major reorganization to embrace states from both sides of the Cold War – arguably with serious adverse consequences to itself (Stirk, 1996, pp. 258–60). The European Union cautiously invests money and preconditions for membership in Central Europe. Its own mid-1990s expansion to embrace Nordic countries and just one Soviet-bloc country, East Germany, has meanwhile reverberated through its internal political structure. Indeed, each extension into former margins makes the functional organization of the whole more difficult (Miles *et al.*, 1995) – as one might expect when absorbing that which is *ex hypothesi* difficult to accommodate. Furthermore, perpetual debate regarding the future inclusion of other territories continues to generate a range of possible futures, geometries and margins for the EU. Much depends on where future boundaries are set: eastward enlargement, EMU, Schengen, the CAP, the Barcelona process, etc. By taking it for granted that shifting boundaries are determinants in the character of *any* integrated whole, a margins research agenda would expressly accommodate this sort of thing. On these common-sense grounds, then, the margins appearing *in* Europe seem to be increasingly important points to conduct research.

Secondly, there is a naive theoretical reason why we should do more research about margins. The margins can be thought of as those parts whose position in a governed whole is least reliable. Following this simple manner of thought, we could suppose that interactions between centre and margins would be difficult for, and therefore would have a significant impact on, both. Furthermore, a straightforward theory of the integration would anticipate that, where territorial extent is increasing, the margins of integration would over time become progressively more significant. Such naive theoretical thinking has more to offer. In any given space of governance, one or more locations or institutions would aspire to be the ordering focal point of the whole: that is to say, it or they would seek to establish terms in which to dominate the rest, in which they were the 'centre(s)'. We could plausibly go on to suggest that the larger a territory, the more extended the borders, the more scope for aspiring centres and for marginal territories that hinder their centrality. Thus, the larger the extent of integration, the more significant – other things being equal – the impact of relations between putative margins and a putative centre or centres. A surprisingly revealing historical parallel can be found in the way Rome compromised with its subjects in the furthest-flung territories (Mattingly, 1992). In short, with expansion the margins of *any* extending system of governance would increasingly be points with significance for the whole, and therefore key points to research.

Following this same, albeit rather pat, reasoning, one can derive two further points. The margins of any field of governance extending in a *non*-territorial sense should possess the same significance as for a territorially expanding one. That is to say, *metaphorical* margins, i.e. groups, or areas of activity which are least under control, could be determinant for the whole just as much as marginal territories could be. One thinks of the 'war on drugs', or organized French farmers, or holders of foreign exchange in the City of London: in each case the integrated system has to undergo real change to extend its governance over the activity or group in question.

Furthermore, there is the matter of research perspective. Evidently, a centre–margin relationship can be looked at either from the centre or from the margin: neither has necessary logical priority. Yet it is the case that most research chooses the perspective from the centre outward. There are no doubt many reasons for that: including the fact that there are fewer centres than margins and that centres tend to be metropolitan places where intellectual and cultural power sets the agenda. Yet as the extent of the integrated order pushes forward, both territorially and metaphorically, margins can be expected to increase both in number and in significance. The centre view accordingly becomes more and more partial.

The third and final reason for pursuing margins research is, of course, the entry of globalization into everyone's research agenda. The exact extent of the manifestations of 'globalization' are easy to dispute. But however epochal it may or may not be, globalization challenges the previous assumptions about the relationship between autonomous processes and the overall, political organization of social life. In particular, the dominant political instance of recent centuries, the national state, no longer appears the obviously optimal level for the exercise of political power over the economic sphere. Some believe that the outcome of integration in Europe will be to replace the old national states with a new instance, more effective because larger. Others believe that the difficulties of the old national states will undermine equally *any* supranational surrogate: 'The state has become a globally extended sphere of meaningful activities. Multiple agencies engage constantly in defining the nature and limits of their respective jurisdictions' (Albrow, 1996, p. 64). In the context of globalization, 'integration' may be either a coalescence of formerly discrete territories or the obsolescence of boundaries altogether. Whichever way one sees it, however, the relationship between instances of governance and their margins again emerges as a crucial problematic.

Margins in European integration

In this volume, then, we take up some of the possibilities in a new view of the changing boundaries of Europe. Widening our view from the fixing of borders to the impact of margins, we ask: How might what happens at the edges alter the overall character of the 'integrated' Europe itself? In other words, we envisage the paradoxical possibility that Europe may be shaped by what happens at its 'margins'.

What makes margins marginal?

Part I pursues in greater depth the character of margins and what makes them. The starting point, in the first chapter, is an examination and defence of distinctions between borders, boundaries, edges, margins and so on. The chapter then explores ways to conceptualize the effects of margins upon European integration, in particular the potential impacts of the marginal on the whole. While none of us would suggest that it is the margins alone that determine the overall character of integrated Europe, we do maintain that crucial effects can be traced to interactions with the margins – enough to speak sometimes, as does Helen Hartnell, of 'initiative and innovation . . . found in those remote, off-centre, in-between, marginal sites'. Accordingly, the first chapter sketches out how such powers and impacts are possible. It argues, furthermore, that both political awareness and theories of integration have responded to circumstances by moving to accommodate to the power of margins. Whilst the attempt to use nothing other than a 'marginal geometry' (Parker, 1998) to describe and account for integration would be misguided, the studies in this book justify the project of looking afresh at integration from the margins instead of from the centre.

Other chapters in this first part offer other takes on the nature and role of the marginal in European integration. Helen Hartnell's is both a case study and a conceptual exploration, for her knowledge of Eastern Europe is combined with a commitment to overcome the perspective which sees the EU's relations with outsiders only from the point of view of those at the centre of the EU. Hartnell steps outside that 'telescopic, distancing vision of integration', to identify various tensions, or 'pathologies', built into the position of Central and Eastern European 'outsiders'. She then sets out the strategic possibilities of their situation and their scope for impacting on the EU and integration itself.

The remaining chapters bring the perspective of economics to bear on how margins are created. Valerio Lintner reviews (especially for non-economists) the standard accounts of how economic inequalities between territories will decline in open integrated markets. Few economists would go so far as to suggest that borders become entirely 'irrelevant' (Nordhaus *et al.*, 1991; Selm, 1997), and propose that in contemporary open markets differences across boundaries might gradually dissolve (but see Chisholm, 1995). Lintner's chapter, on the other hand, sets out the conceptual framework which has long led to the expectation that inequalities will decline. That in turn informs the anxiety, or lack of it, that economists and policy-makers display when it comes to managing economic differences between territories. Combining politico-economic theory with case-study material, Leslie Budd then takes a look at the European Union's policy efforts to date to reverse perceived economic distortions between the centre and the periphery. His conclusions are sceptical: in the new global environment, integration does more to let loose economic forces compounding the inequalities than it does to reverse them. If a new version of neo-colonial economic and social relationships is to be avoided (both across what Russell King (1998) calls the new 'migration' frontier of Europe and within European society itself), he argues that much more will have to be done by the Union, especially on its Mediterranean margin.

By the end of the first part, then, a case has been established for various possibilities in events and actions at the margins of Europe: to confront, to demand, to manoeuvre, or merely to differ in ways that can shift or unsettle the internal balance of integrated Europe as a whole. The remainder of the book looks at particular instances, dividing the possibilities loosely into the economic and the political.

Trade and governance at the margins

The first area where we look more closely for effects and tensions on the margin of integrated society in Europe has always been the prime objective of integration. The integration of markets stimulates a number of economic processes that can create or amend the margins: differential distribution of factors of production, distinctive patterns of wealth or productivity and so on. The substantial extension of market integration contained in the Single Market programme of the 1980s thus stimulated debate and policy initiatives to manage these effects on the margin (Bulmer and Scott,

1994). What is harder, of course, is to say exactly what these effects are and how – or indeed whether – they can be politically managed.

Mina Toksoz considers the margin of Europe which Leslie Budd's chapter identifies as the most problematic. She tells the story of the European Community's efforts to manage trading, and other relations with the countries of the Mediterranean. These have gone through a number of stages, none of them particularly successful in the face of both regional and global political dynamics: the managed exchanges of the 1960s, the upheavals of the 1970s, the relaunch at Barcelona. This last has since been much frustrated by divergencies between Mediterranean countries, by the geopolitics in the Middle East and by recent global economic shifts. In the Mediterranean, we find an inveterately troubled relationship between Europe and one of its margins.

Chris Flockton returns our attention to the margin that is uppermost in the EU's own debates, Eastern Europe. *Prima facie* Eastern Europe compounds the difficulties at the Mediterranean margin, notably by redirecting investment. But Flockton's detailed analysis of the different flows and forms of direct investment in Central and Eastern Europe suggests a surprising conclusion. Political common sense would suggest that enlargement would simply reorder the relative prominence of the margins of the Union, switching attention and benefits from a Southern to an Eastern periphery. Carefully analysed, however, a quite different picture emerges from the evidence available. Eastward expansion alters, if anything, the economic structure of the present economic *core* in northwestern Europe. Conversely, Flockton finds the impact of this new investment on the structure of the economies of the Central European margin itself to be an ambiguous, partly beneficial mix.

Ann Kennard's case-study of evolving organization in the same geographical area focuses attention on the interface between the economic and the political. The boundary between Poland and reunited Germany is a particularly entrenched marginal territory of Western Europe. Kennard looks at the practices that have evolved to manage developments in that border territory. Whilst she shows how much cross-border organization is being undertaken, she also indicates ingrained impediments that mean it will be some time before this frontier of Western Europe is shifted altogether.

The politics of marginality

An important source of significance for marginality, in European integration or elsewhere, is the scope that it offers for the pursuit of power via the deployment of inherently contestable claims to position. To put the point another way, being on the margin is *political*: it implies particular strengths and weaknesses for those engaged in contests for power and influence, and they react by constructing, challenging, or *re*constructing their own marginality.

José Magone points up the impact of lessons in the evolution of governance on the southern edge of the European Union. Surprisingly for those who see the centre of political innovation in the North, Magone argues that Iberia has been a proving ground for new ways of combining the different rationalities in policy formulated at different levels of power – 'multi-level governance', as it is sometimes referred to. Thus, in Spain and Portugal, the management of marginal territories and marginal interests has been reorganized, putting those countries at the centre of a development of European democracy which, for Magone, is better adapted to accommodate margins than those elsewhere in Europe.

Christopher Flood's chapter takes a striking example of *using* claimed marginality: the way that the political right in the UK has declared Britain to be marginal, different and put-upon by Europeans on the mainland. Yet by doing this they actually reassert a British *superiority* over Europe, and hence magnify the impression of Britain's freedom to act independently of the rest of the Union. At times this Eurosceptical position even manages to invert Britain's marginality and resituate the country, at least in the minds of the Eurosceptic, at the centre of the globe.

While British Eurosceptics have ingeniously turned British marginality on its head, Czech culture and self-identity seem to have overplayed a comparable hand. A one-time cultural centre to rival many further west, the Czech Republic started out boldly asserting its *right* to 'rejoin' a political and cultural family which it believed it belonged to from ages past. The meeting of that historical view of the world with the inertia and the narrow, hard-nosed economic or institutional demands from the EU side has, on the other hand, mellowed the Czechs' sense of their place. After years spent knocking on the door, historical cynicism has increasingly surfaced, mixed with mutual recriminations over ignorance and political ineptitude on the part of political leaders.

Finally, Hugo Frey takes the discussion into a different dimension, the politics of historical memory: that is, shared interpretations of the past which often discreetly inform a group's response to the present, legitimizing or challenging political positions. Conventional wisdom would hold that European integration should be backed by a common European historical memory. Failing that, integration might be possible if divergent historical memories could merely be suppressed. Frey challenges both that expectation of unity, and a widely believed alternative explanation of things, according to which, in order for integration to happen in the first place, post-war Europe suppressed all historical memory – temporarily, of course. Integration, argues Frey, has always had to accommodate divergent historical memories which construct different 'Europes' in the past.

As can be appreciated from the synopses above, rather than finding – or expecting to find – a definitive overview of Europe and its integration, this book explores the possibilities within an amended research agenda. According to that agenda, marginality is a common feature of politico-economic entities – perhaps inevitable, for some even welcome. We can expect it to be more and more in evidence in Europe, and progressively more relevant to understanding European integration. The various chapters in this symposium – be they theoretical reflections, case-studies or a combination of the two – establish the potential for research on Europe from the point of view of the margins. Such research will not replace the established agendas of institutional, historical and political science study on European integration; but, by putting firmly in question the assumptions of others as to the underlying unity of the 'Europe' being studied, it does suggest that which may be seen only in a distorted light by following other approaches: the impact of increasing marginality on the character of Europe viewed as a whole.

References

Albrow, M. (1996) *The Global Age: State and Society Beyond Modernity* (Cambridge: Polity).

Anderson, M. (1996) *Frontiers: Territory and State Formation in the Modern World* (Cambridge: Polity).

Bulmer, S. and Scott, A. (eds) (1994) *Economic and Political Integration in Europe: Internal Dynamics and Global Context* (Oxford: Blackwell).

Chisholm, M. (1995) *Britain on the Edge of Europe* (London: Routledge).

King, Russell (1998) 'The Mediterranean: Europe's Rio Grande', in M. Anderson and E. Bort (eds), *The Frontiers of Europe* (London: Pinter), pp. 109–34.

Mattingly, D. (1992) 'War and Peace in Roman North Africa', in R. B. Ferguson and N. L. Whitehead (eds), *War in the Tribal Zone: Expanding States and Indigenous Warfare* (Santa Fe, NM: School of American Research Press), pp. 31–60.
Miles, L., Redmond, J. and Schwok, R. (1995) 'Integration Theory and the Enlargement of the European Union', in C. Rhodes and S. Mazey (eds), *The State of the European Union*, vol. 3 (Boulder, CO and Harlow: Lynne Reinner/Longman), pp. 177–94.
Nordhaus, W., Peck, M. and Richardson, T. (1991) 'Do Borders Matter? Soviet Economic Reforms after the Coup', *Brooking Papers on Economic Activity*, no. 2, pp. 321–40.
Østergård, U. (1998) *Europa: Identitet øg Identitetspolitik* (Copenhagen: Munksgaard/Rosinante).
Parker, Noel (1998) *Geometries of Integration*, DUPI Working Paper No. 1998/15 (Copenhagen: Dansk Udenrigspolitisk Institut).
Selm, B. van (1997) *The Economics of the Soviet Break-up* (London: Routledge).
Stirk, P. M. (1996) *A History of European Integration since 1914* (London: Pinter).

Notes on the Contributors

Bill Armstrong is a one-time University of Surrey research scholar and freelance writer.

Leslie Budd is Senior Lecturer in Economics at London Guildhall University.

Peter Bugge is Associate Professor in Czech and European Studies at the Department of Slavonic Studies, University of Aarhus, Denmark.

Christopher Flockton is Professor of European Economic Studies, Department of Linguistic and International Studies, University of Surrey.

Christopher G. Flood is Head of European Studies, Department of Linguistic and International Studies, University of Surrey.

Hugo Frey is a Lecturer in History at University College Chichester, and was formerly a University of Surrey research scholar.

Helen E. Hartnell, JD, is Associate Professor at Golden Gate University School of Law in San Francisco, California.

Ann Kennard is Senior Lecturer in German in the Faculty of Languages and European Studies, University of the West of England, Bristol.

Valerio Lintner is a Reader in European Economics at London Guildhall University.

José Magone is Lecturer in European Politics, University of Hull.

Noel Parker is Senior Lecturer in European Politics, Department of Linguistic and International Studies, University of Surrey.

Mina Toksoz is senior strategist for emerging Europe at ABN-AMRO Equities Ltd (London).

Part I

What Makes the Margins Marginal?

1
Integrated Europe and its 'Margins': Action and Reaction

Noel Parker

Is Europe getting edgier? Outsides, boundaries, borders, frontiers, peripheries, margins

'Is Europe getting edgier?' is not a merely facetious question. Someone is 'edgy' when unable to settle, unstable, not confidently positioned in the world, trying to avoid falling, on the look-out for unexpected dangers. Over recent years one can easily observe that 'Europe' is edgy in this sense. The European Union, NATO, the Council of Europe, etc., are preoccupied with the problems of organizing relations with, or integrating, various forces at the edges of their authority. The absorption or not of the countries of Central Europe; the establishment of a stable peace in countries further to the east; promoting a new democratic market identity for Russia, where relations can be beneficial rather than threatening to Western Europe; the management of commercial relations with countries in the Mediterranean south; holding back the fall-out from their internal dissent or disintegration: all these are instances of anxiety about unexpected impacts from what lies on the edge of Europe. They appear on the current agenda of European bodies, and on that of this book.

It is not too much to add that the principal occasion of these preoccupations is the transformation of 'Europe's' pre-existing edge. Whatever one felt about what lay beyond the Cold War border of Western Europe, there was a well-established edge, and a stable relationship with the West, marked by clear organizational and ideological differences – and sustained through the deployment of quantities of global power. Once these edges had been firmly established in the 1950s, 'Europe' could afford not be edgy about itself: its outside was clearly identifiable, manageable and protected. Arguably, that Cold War edge was irreplaceable as a firm basis for the identity of Europe. With that certainty gone, Europe has become 'edgy': preoccupied with how to formulate and manage its borders, and what to do about its 'outside world'. Jonathan Story has given a particularly apocalyptic version of Europe's post-Cold-War malaise precisely in terms of its managing those boundaries (1997, pp. 41–2):[1]

> the fragmentation from the peripheries of Europe [is] quite capable of entering the core, destroying the achievements of the past four decades.... The real Europe is once again pregnant with conflicts... the EU of concentric circles, inner groups and many speeds is merely an older and differentiated Europe in a new guise.

If we look at supra-national European politics not only in the EU but also in NATO, the Western European Union (WEU), CSCE, PfP, etc., we find expressions of this: so many different tactics to accommodate one or other of the newly adrift edges of Europe.

At the same time, new political formulas are surfacing to cover what one might refer to as the '*inside* edge', the awkward parts of the European Union that are formally already on the inside, but will not accord with the dominant direction of movement. Internal relations with certain more or less established member states (MSs) on the edges of the existing Union have long been of concern. I have the UK in mind, of course, but likewise Denmark, or Greece, or Spain and France with their different southerly-leaning agendas. These also figure in the thinking behind this book. To respond to the problem of inside edges, negotiators have inserted into treaties formulations such as 'subsidiarity', the 'opt-out', 'unanimity minus one', 'constructive abstention' (Stubb, forthcoming). Likewise, Union politicians have spawned a number of new political formulas, such as 'flexible integration', 'two-speed Europe', 'variable geometry' and so on. Some commentators take these new political formulas very seriously, arguing that they mark a new phase in the *form* taken by integration – sometimes seeing them as the thin end of a wedge which, unless checked, will undermine the coherence of the EU as a whole (Duff, 1997). For myself, these notions appear more as political window-dressing to paper over the fact that different forces want the European Union to take different directions. But whatever the truth of the matter, the appearance of formulas such as these certainly points to a European politics that has to deal with edges which are more strongly felt than before.

A broader setting for these uncertainties about the edges in Europe may be found in a current 'deterritorialization' of the sociopolitical order of component states themselves. In that light, European integration can be seen as, at one and the same time, victim, agent and antidote to the wider phenomena of globalization. For a commentator such as Jean-Marie Guéhenno (1995), the devaluation of national-state borders has the effect of devolving the exercise of power in chaotic ways that can undermine Western Europeans' existing framework for democratic politics. In such a situation the European Union's capacity (as yet largely latent) to coordinate political expressions at a supranational level may be the saviour of democratic practices that will otherwise be hollowed out by what Guéhenno refers to as 'lebanonization'. In a study that brings political anthropology to bear on questions about the policy priorities of the EU, O'Dowd, Wilson *et al.* (1996) examine ways that the softening of borders has given scope to cultures in the border*lands* which challenge the seeming loyalty to national states. These borderland reactions are crucial forces in the general decline of the marriage of statehood and nationhood. In some ways this is a development waiting to happen, since nation-states' defining borders have everywhere been rooted in coercive exclusion or inclusion and then sanctified from above. Insofar as it has lent authority to MS boundaries, European integration has been an antidote to the effects of weakening of borders. But it has also been a prominent conduit of that weakening itself: as an authority behind free movements across pre-existing borders; as a policy centre contradicting the MSs' capacity to shape infrastructure as they wished; as a patron of political processes and identities at the subnational level; and as a proponent of *rival* borders implying – even if unsuccessfully – an alternative loyalty.

In a broader historical context we might simply suppose that the problem of uncertainty over the edges of sociopolitical orders is not so much new as newly resurfacing. The separation of discrete national-state orders is, after all, fairly new in historical terms and hardly stable (Tilly, 1992, pp. 45–7). The long history of laying and re-laying frontiers in Europe simply continues today (Anderson, 1996, pp. 37–76). And national states themselves have always circled in and out of larger Europe-wide arrangements. In or out, the status of the UK has *always* been an issue *within* Europe: from the nineteenth century, when it was the hands-off counterweight in the European balance of power; to the 1940s when it could have been the economic cornerstone of integration; to the 1980s when its structural distinctness produced the 'anomaly' of British contribution to EC budgets and almost derailed the internal workings of the Council of Ministers; to the early 1990s when the home of the biggest financial market in Europe joined, and was then dramatically forced to leave the ERM/EMU project. Similar observations could be made about the position of Denmark, as a state which is accustomed to being preserved in between, and with the accord of, larger powers and which therefore knows how to use its position to preserve a semi-detached distinctiveness – for example on rights to land ownership. Even Germany, the EU's quintessential insider, has a past as Europe's most difficult outsider, taking its place late in the European political system and distorting or disrupting it in the process. Its eccentric ethnic principle of citizenship remains as one relic of its long efforts to form stable territorial boundaries, both internal and external.

Thus there may be another way of looking at the new importance of Europe's edges, which I would like to pursue over the course of this introductory chapter. It is the thought that, like those of any political order, the edges of integrated Europe are *never* really stable. If Europe currently shows signs of heightened awareness of that fact, that is because boundedness is more difficult to take for granted at some times than at others. Though in reality the edges in, and of, Europe are always contestable, the present moment is one when they have to be more openly considered than at any time in the recent past. For various reasons, political actors in Europe are increasingly grasping at concepts intended to embrace, and manage as best they can, the diverse margins within a partially integrated Europe. It may be the end of the Cold War, or the European Union's increasing size, or the increasing range of governmental and political areas that the Union has to try to take an interest in. It may be globalization reversing the previous fixity. It may simply be a growing cussedness on the part of the member states and civil society. However that may be, I suggest that the recent 'edginess' points us towards something that has always been more or less true: that Europe cannot expect the luxury of clear-cut edges. It has to expect to be a number of shifting, overlapping, interlocking entities which create, and interact as *margins*. Two points follow from that. First, edges should never be regarded as simple absolutes. Secondly, European integration is a spatial matter, which we can expect to vary according to location in its completeness, its depth, and even its form.

We have to discriminate between different *kinds* of edge, both outside and inside Europe's integrated areas. To explicate further the generic term 'edge', I suggest a series of cognate terms: 'boundaries', 'borders', 'frontiers', 'peripheries', 'margins'. This series implies a progressively more substantive view of what lies *at and beyond* the edge. Whereas the 'boundary', for example, is primarily the point where the present territory stops, the 'border' is what has to be crossed to get to the adjacent one. The 'frontier' actually seems to *require* that something be done with regard to what lies beyond the edge: it may be that defence is needed against the threat from the other side; conversely it may be that – as in the image of space as 'the final frontier' – the frontier must be breached and taken forward (Anderson, 1996, pp. 9–10). With 'periphery', an altogether more complex relationship is implied with that which lies beyond. The territory beyond is *already* in a definite relationship with this territory. Furthermore, peripheries (and margins) are substantive territories in their own right. Certainly, their character as peripheries (and margins) might be defined in terms of the area to which they are peripheral/marginal; but they undergo processes and exhibit features over and above those of the mere passage of others *through* them. They are not merely the product of external powers. Hence the relationship of a periphery, or a margin, to the whole is more likely to be a *two-way* relationship. In the case of a periphery that relationship is implicitly a subordinate one. But for margins, that may well not be the case.

Margins and wholes

The expression 'margin' implies the most complex of relationships between what is within and what lies outside. There may be movements across the margin or, more probably, movements from *within* it into the territory of which it is the margin or out to the surrounding area. Like a periphery, a margin is itself inherently difficult to pin down; it may grow, shrink or move altogether. But what we refer to as a margin has, I suggest, greater autonomy than does a periphery *vis-à-vis* the territory to which it is marginal. Etymologically, the expression is related to the old word 'marches', those edges that were difficult to penetrate and from whence various shadowy dangers threatened a feudal order.[2] Thus the impact of the margin may itself reconfigure the edge, a possibility not included in the meaning of any of the other terms. Hence, the term 'margin' was adopted for these studies, which focus on the spatial differentiation in integration and on the impact which that differentiation has upon integrated Europe.

The choice of 'margin' in the title of a book that examines the edges of Europe was prompted, then, by a desire to focus on ways in which its edges may be impacting upon Europe as a whole. Alone amongst the above expressions for the edge, it is margins that may 'bite back'. The margins of a given order are, one may say, an edge which may plausibly be incorporated as part of the whole or not. And that indeterminacy gives to margins a greater impact than other kinds of edge: the possibility that they might or might not be joined, or might leave the whole impacts upon the whole itself. That is why action 'on' or 'from the margin' offers quite particular leverage and room to manoeuvre. The margin is harder to suppress from the centre; it can exploit the distinctiveness and autonomy preserved at the edge; it can threaten secession and/or, if outside, external rivalry or attack. To deal with a margin, then, presents particular problems to the given order as a whole. If the margin is incorporated into the whole (or leaves it) the overall structure is altered; if it remains on the outside, an awkward external relationship persists; if it is incorporated, an awkward *internal* relationship appears. Hence, paradoxically, the margin may at times be observed defining the character of the whole.

Being marginal is often unthinkingly taken as a negative. But, though marginality as here defined certainly does present a peculiar set of problems to a sociopolitical order as a whole, one strength of the conception we pursue is to dissociate marginality from the idea of *inferiority* to, or *dependence* upon, a corresponding core. Certainly, dependent relations to a 'core' area can easily exist, as with a periphery, and to that extent parts of this book examine the dynamics of core–periphery relations. But the relationship of the margin to the whole embraces those of core to periphery as a subcategory. Logically speaking, there do not have to be relations of dependence between margin and whole, and any dependence there is may not be significant. In short, margins may bite back in a way that mere peripheries cannot.

If one thinks about its edges in the terms we are following in this book, many areas could show up as margins of the European Union. Confining the discussion, for the sake of simplicity, to whole national territories, Greece is marginal because it joined late and only has land links with non-EU territories; the same applies to Sweden and Finland (except that they have a border with each other); Norway keeps *not* joining; the UK has no land links (apart from the politically peculiar border with the Irish Republic), joined late, and often shows signs of wanting to leave again. The distinctive half-in-half-out position of each of these marginal countries contributes to the form of European integration overall. In its pursuit of regional rivalries, for example, Greece hardens the EU's Eastern boundary, insisting that Turkey must be left on the outside, that a Greek Cyprus should be allowed in, and that FYROM (the Former Yugoslav Republic of Macedonia) does not properly exist at all. For their part, the Scandinavian countries bring to the EU their values in matters of environmental and social standards (which were developed in their late, Scandinavian experience of industrialization) as well as their awareness of the economic integrity of the Baltic Sea, and the consequent priority to be given to the EU's embracing new 'Baltic states'. Conversely, Norway's *absence* from the EU sustains the peculiar structure of EU–EEA–EFTA (to which Switzerland's long-standing marginality has likewise contributed), and limits the effectiveness of the Common Fisheries Policy. The UK stands in the way of full integration of Schengen into the EU structure. More profoundly, its version of state sovereignty, formed through its past as a free-booting global power on the maritime edge of Europe (Clark, 1991; Gamble, 1991), greatly reinforces the intergovernmental character of the EU and stands as a political obstacle to EMU membership (Risse *et al.*, 1998). Likewise, its perpetual suspicion of mainland 'European' standards of social protection (in turn grounded in economic necessities felt by a de-industrializing post-imperial state on the margin of the later industrial centre of Europe) limits the integrity and viability of what is called 'Social Europe'. For better or worse, in these and other instances the margins are shaping European integration.

The common impact of those countries' marginality on the EU is to alter the extent and the nature of the EU as a system of integrated governance. Territorial margins, I suggest, are those areas embraced only tentatively by a structure of governance, such that either presence or absence would entail some change in the overall entity. In each case there is, as it were, a price to be paid for the integration of the margin into the whole. This yields two further observations, which indicate how widespread the tensions with marginality may be in structures of governance, integrated or otherwise. First, it would be both logically sound and empirically accurate to remove the restriction to *territorial* margins and include margins in a metaphorical sense: marginal *groups* or marginal *activities*. Margins, one might then say, are those areas of social activity more difficult to embrace under a structure of governance, such that accommodating, altering or suppressing them entails a change in the system overall. Secondly, governmental orders have *always* had to deal with the problem of how to accommodate margins in both the territorial and the non-territorial sense.

In this, the EU, which also aspires to be some kind of governmental order, merely follows centuries of national-state experience with margins both territorial and non-territorial. Many historical studies support this contention. From the strategies of the high-medieval expansion into north-central Europe (Bartlett, 1993), to the tellingly named 'Councils of the *Marches*' which in the sixteenth century imposed order on Wales and the North of England, to the peculiar status of *pays d'élection* which brought regions further south under the authority of the French crown in Paris, and so on and so on: to secure their territorial hold, political orders have always had to suppress, accommodate and/or negotiate with forces at their margins. Studies combining history and sociology support the inclusion of *activities* as well as territories within this claim. Eugen Weber (1976) and Charles Tilly (1995), for example, show respectively the French and the British states incorporating their marginal, rural populations under the central, national political order, not least through extensive military mobilization. European national states have always needed strategies to quell the centrifugal impulses of their peasantry and/or engage its loyalty. Charles Tilly (1992) has attempted a general account of what states have had to do to incorporate another growing form of social *activity*, urban commerce, and the *groups* engaged in it – though in practice they have never been able to contain them entirely. Capital exhibited a secular tendency to autonomy which at best forced liberal concessions from developing states, and at worst crippled them. Rosecrance (1986) has even argued that, in the twentieth century, the *increasing* independence of that 'trading world' is replacing the 'military–political world' and recasting the form of the state in the process. The history of the EU's relationship with international capital surely offers abundant parallels, culminating in the massive gamble of EMU: an attempt to invent a new monetary structure that will win over global business, industry and finance which MSs have progressively released from political control (Tsoukalis, 1996). At the time of writing (and no doubt for some years to come) the EU also sticks its toe into the waters of harmonized corporate taxation – a drama that readily brings to mind Tilly's saga of states teetering between coercing and appeasing semi-autonomous commerce.

These, then, are the grounds that provide the analytical importance of margins in European integration. I observed earlier that much of EU politics has always been about the problem of accommodating margins. Like other sociopolitical entities the EU has always been confronted with a need to wean both spatial and social margins from their possible separateness, fit them into some kind of single structure and obtain their loyalty to the integrated Europe. EC agricultural policy has been a striking, and highly expensive, strategy along those lines: an arrangement intended to embrace marginal, rural society under the governmental institutions of Europe by protecting those who live on the land against the vagaries of the weather and the world market. Italian membership of the EC was from the very start a way of drawing closer to the north-western core of Europe a country that was both geographically and metaphorically marginal. In the non-locational sense Germany has been a marginal country, a political pariah needing to be incorporated if it were not to cause more trouble to its neighbours from outside (Breuilly, 1998). And the effects of the danger that others perceived in Germany's sense of its capacity to go its own way economically are plainly visible even today – for example, in concessions made to German financial priorities and structures in order to lock her into the monetary integration of Western Europe.

What is different at the margins?

Because action 'on' or 'from the margin' offers particular leverage and room to manoeuvre, we can expect that actors placed there would behave in characteristic ways. I earlier defined the marginal as areas tentatively embraced by a structure, such that their presence or absence would entail a change in the whole. A. O. Hirschman's classic account (1970) of how leaving affects an organization may offer key concepts which apply, albeit with amendment, in the context of the European order as we are considering it here. Hirschman counterposes 'exiting' from a body to 'voicing' complaint,[3] and sets out circumstances where one or the other is more likely in itself and/or more effective in altering the body in question. What he calls 'loyalty' enters the picture as a retardant on exit (and on re-entry), particularly in the case of bodies such as states, families, or organizations providing public goods. For public goods are enjoyed, or endured, by everyone regardless of whether they contribute or have a say in their provision. Hence it is peculiarly hard to extricate oneself from the supply of public goods. There is a clear parallel with European integration, which affects unwilling participants and peripheral non-participants as much as those who are fully committed. Effective exit is difficult; so one should expect loyalty. Hirschman contends that loyalty gives a pejorative value to exit, but also reinforces the shock of it (pp. 82–3). Therefore loyalty increases the effect of threatening exit.

> That even the most loyal member can exit is often an important part of his bargaining power *vis-à-vis* the organization. . . . [I]t appears that the *effectiveness* of the voice mechanism is strengthened by the possibility of exit. The willingness to develop and use the voice mechanism is reduced by exit, but the ability to use it with effect is increased by it.

At first sight this appears to resemble the behaviour of the margins, which are by my definition in a position to exit and able to use the threat of doing so. The margins, one might say, are outsiders/exiters in principle.

Yet one of Hirschman's findings flies in the face of the reasoning here. He argues that it is the *most loyal* – rather than those *recognized as marginal* – who make the greatest impact through threatening to leave. Elsewhere, again, Hirschman argues that it is core voters, with 'nowhere else to go', who raise their voices loudest and hence 'voters at the extreme [who are over-represented in the party's active supporters] manage to inflict actual vote losses on the party if it moves too far to the center' (p. 73). As to this last point, it can only be said that it does not accord with contemporary political experience of parties scrabbling to occupy 'the centre ground'. Hirschman appears to understate considerably the fear in an organization at the relatively probable loss of the *less* loyal. For the organization's fear of loss is the product of the members' potential to complain *and* the likelihood of their departure. While those who are locked in have a strong motive to complain, they are in reality unlikely to leave; but those who do have an alternative outside can make a more realistic threat to depart.

To carry over Hirschman's concepts to European integration, we have thus to reverse some of his findings. In the club of European integration the 'membership' is not so numerous that departures can be easily sloughed off by the organization as a whole. Or to make the point from another direction, ex-members or non-members do not simply disappear from view, ceasing to impact on the body as a whole. Thus the situation as Hirschman understands it for public goods may not obtain. If part of the organization left, we would not simply have the same effects falling on members and non-members, with those outside getting the same provision without the influence of insiders. Rather, the organization itself would be altered. It is as likely to suffer from the outsiders as the outsiders are to suffer from the public goods/ills generated by the organization in question. Hence, loyalty is less strong a pull for the members than in the classic public-goods situation. Yet, for the majority firmly on the inside of integration, loyalty does still apply – either as a value in itself, or in the sense of there being nowhere else to go. By comparison with the established, 'loyal' members, the margins have leverage because they are plausibly able to leave (or resist joining) and because, by sustaining an awareness of their own, separate identity, they *reverse* the value Hirschman gives to loyalty. Against the background of others loyalty/enclosure in integration, the margins have leverage because they are outsiders/exiters in principle in a situation where loyalty applies to the rest.

Having specified the particular mode of influence at the margin, we can turn to the question of what makes a margin marginal. What I have said up to now about the scope for areas, activities and actors that are marginal could make anything a margin. Perhaps we should be willing to go some way in that direction and accept that the increasing predominance of edges in the perceived reality of European integration is making it possible that *all* its component parts are a margin in relation to something or other.[4] This may express an underlying reality of integration that is becoming more evident at the present time. Conversely, there is a risk of truism in my earlier definition, contained in the formula that a margin 'may plausibly be incorporated as part of the whole or not'. For there is no *a priori* limit to what can 'plausibly' be inside or outside: hence, everything could be included under the concept, and the notion of marginality becomes vacuous.

There is no clear-cut, once-for-all test of what is marginal. On the other hand, there are effective limits which reflect dominant, understandable interpretations of the situation. We can see that from the way that different countries may be seen as 'plausibly' in or out under different conjunctures. In the first decade after World War II, Benelux could plausibly go into integration or not; now it cannot do other than remain inside. Up to the present, the UK has been seen as plausibly in or out, though perhaps to a diminishing degree. In the years immediately after 1989 it suddenly became plausible to some that Germany could go its own way: the opening of its eastward limits had freed it from the overriding necessity of seeking to accommodate its western neighbours. In the years after 1945 also, some saw Germany as potentially in or out of any integration: but at that time the decision lay with those *outside* Germany, whereas more recently it lay with Germany itself. Post-1945, French politics perhaps did more than anything else to make German exclusion plausible; while US strategy initially did most to counter that, followed in due course by Germany's own economic resurgence. Amongst the studies in this book, we likewise find instances of marginality attributed and reversed (Iberia), marginality imposed and denied (Poland, the Czech Republic), and marginality embraced (Britain, the ex-USSR). For being marginal is an effect of human actions on various levels, perhaps even a *choice* brought about by a wide range of conditions. It has specific attributes, possibilities and limitations.[5] Some make that choice for themselves; some have it thrust upon them; some seek to *navigate*,[6] or even to *use* their marginality; others try to unmake it.

Margins and integration theory

Indeterminacy in applying the concept of marginality should not necessarily discourage researchers and analysts of European integration. For insights expressed in the growth of deconstruction and postmodernism have some force in legitimizing indeterminacy. In political science and international relations, such ideas have attracted reactions ranging from hostility, to guarded interest, to whimsical enthusiasm for the new narrational possibilities (e.g., Rhodes, 1996; O'Sullivan, 1993; Bleiker, 1997, respectively). Yet the central claim of those philosophical tendencies remains suggestive: that it is a mistake to determine identities on the basis of some single, unitary essence which is supposed to contain their inherent structures. Thus, deconstructionism has been applied, for example, to redefine nationhood so as to incorporate within it an interdependence with that which lies *outside* the nation (Kristeva, 1993).[7] In the same vein, deconstructionist reflections on the identity of Europe have highlighted the conceptual difficulties implicit in conceiving Europe as a *singular* identity (Derrida, 1992; Lenoble and Nicole, 1992). Philosophically speaking, these findings make common cause with newly developed positions in integration theory founded on evidence for the dissolution of clear-cut political entities in the new integrated Europe – notably the concept of 'multi-level governance' operating through a plethora of regional authorities, policy-networks and political arenas that bypass older, fixed sovereign territories (Christiansen, 1997; Marks *et al.*, 1996). Right or wrong, these theoretical developments suggest that indeterminacy can be recognized and understood.

The deconstructionist philosophy has been applied specifically to the indeterminacy of borders in a series of lectures by Geoffrey Bennington (1989–92). These pursue Frege, Kant, Hegel and Wittgenstein through their attempts to establish some point at which frontiers can be definitively determined. But whether it be the state, the concept of the ideal, human knowledge as a whole or what, attempts to fix the frontier will fail because, broadly speaking, determining a boundary always presupposes that what lies beyond can be shown to be categorically distinct from that which is on the inside. To take only Hegel, the most explicitly political of Bennington's chosen thinkers, the indeterminability of frontiers can be traced in the inveterate tension between absolutes (Reason, History and so on) and the particularity of the given state's political order or historical circumstances. Insofar as they are accidents of history, frontiers are thus contingent and cannot provide entities such as states with an assured, rational identity or a position in an *absolute* History. But this is only the most *explicitly* political point in Bennington's discussion. Bennington deduces that *all* frontiers are 'political'. For all have to be sustained with difficulty,[8] that is by the play of human powers and intentions. Hence, frontiers are 'duplicitous', 'identity is fraudulent . . . there's nothing like identity' (p. 15).

That sort of thinking suggests indeed that the margins, which are the places where indeterminacy is played out, should not be merely a permissible topic theoretically speaking, but can also be the material of inquiries that promise particular insight. For the problem that even sympathetic theorists in political science and international relations have felt in the postmodernist current has been its tendency to celebrate how indeterminacy *defies* argument and research. Hence demands for a rediscovery of politico-legal forms where difference can be articulated (O'Sullivan, 1993), or for terms of analysis to track the diversification or crystallization of authority (Caporaso, 1996, p. 47f). According to Bennington's findings, at base all frontiers are marginal in the sense that they mark out territory able to join or leave the given entity, so that the certainty of the latter's identity within its boundaries is undermined. The aim of this book is to speak of 'margins', areas that, because they could be on one side or the other of the boundaries defining integrated Europe, challenge its fixity. And the margins – with their specific opportunities, practices and discourses – are an obvious place to look for identifiable forces, rules or practices that might – in European integration as elsewhere – be dissolving or reconstructing political entities.

Researching the margins may then go further than observing blandly that marginality is constantly present and depends upon a lot of possibilities and the choices made within them. A recent study by Mouritzen *et al.* (1996) can point a way forward for that research. It attempts to combine country data on modifications in response to integration pressure with Mouritzen's own spatialization of state behaviour in international relations via the concept of 'environment polarities' (see Mouritzen, 1998).[9] The burden of this concept is that any given state will adapt to/make a trade-off between the forces occurring in its particular location of the power 'constellations' which form in inter-state relations. Put very crudely, the forces in place are a function of the given state's proximity to one or more 'poles' of international power. Mouritzen and his fellow-researchers apply this sort of geometry to a group of countries (Denmark, Netherlands, Finland, Poland, Lithuania, Ukraine) more or less under the pull of European integration during the years 1989–94. That is to say, they adapt an International Relations (IR) concept of poles of state power to a notion that the European Union is a pole – a so-called 'membership pole' (p. 21) – where the prospect of joining or not has impacts on each state's policy values that are comparable to those of inter-state power in earlier IR theory. The topic of their IR-style enquiry appears, then, to be the same as ours: margins of integration, where countries react to the possibility of being members of the integrated whole.

Out of their theoretical mix, Mouritzen *et al.* establish in effect a model for marginal membership of the European Union: a range of political postures for potential members of the EU which vary according to their position in the power constellation formed by the

Union. Taken together, these postures can be assembled in a pattern showing the pull from the EU pole. So far so good. But the results in fact achieved by this approach are limited. The states considered are grouped into three categories: outsiders, insiders and 'would-be insiders'. But only those in the last category exhibit a common tendency in their policy values, namely, to conform to EU norms (p. 273). Others behave heterogeneously, even during this period, which was comparatively favourable for the pull of the EU since membership was widely thought to be the only feasible way forward for states on the margins. As Mouritzen *et al.* are aware (pp. 39f and 299–306), the period from 1989 was quite unusual in concentrating the forces in Europe around a 'unipolar' EU. From 1992, indeed, it already looked less obvious that the EU was the only possible future: bi- or multi-polarity, including the EU, became possibilities again.

Overall, however, the Mouritzen *et al.* study does bear out the feasibility of an approach to marginality in Europe as suggested in the present book. Its conceptualization shows how behaviour might be patterned according to 'proximity' to EU integration. The difficulties encountered suggest, on the other hand, that more possible responses, contrary pulls *and* the exploitation of the situation would all need to be taken into account, and that at the end of the day there would be a large measure of choice about embracing, exploiting or diminishing marginality. Helen Hartnell's Chapter 2 discusses further the divergent pulls felt by state on the eastern margins of the EU in the 1990s. At the very least, there need to be further categories of response,[10] including more or less deliberate half-membership and troublesome non-membership. Both of those were implicit in the earlier discussion of how marginality may be attributed, embraced or rejected. My initial supposition thus remains: that Europe does not have *absolute* frontiers, only relations across frontiers, and that the behaviour of those entities and activities that may 'feasibly' be in or out has constitutive effects on the overall character of integration inside the frontiers of Europe.

The 'Unitarist thesis' in integration theory

To look at integration as the outcome of a concatenation of distinct margin–whole relations marks a considerable shift from the habitual presupposition of political actors in the integrated bodies of Europe, and from a strong predisposition in theories of integration.

International political economy can entertain the theoretical possibility of territorial boundaries disappearing. Geography and anthropology can approach the specific characteristics of border territory, which are functions of human geography regardless of inter-state relations (Minghi, 1991). But for a long time explanations of integration prioritized common features attributed in advance to a Europe which they thus conjured into existence. The theories postulate beforehand firm existential boundaries to that which does not yet, and may never come to, exist: Europe as a single entity. 'Integration theory' has an inherent tendency to build its explanations in this way, on prior common features of a 'Europe' that is already implicitly one.[11] Because they contain that presumption of a pre-existent unity in the outcome of integration, I refer to conceptions of integrated Europe in terms of its commonalities as 'the unitarist thesis'.

There is a common-sense view which comes easily to those involved in, and those analysing the processes of, the EU, to the effect that: (i) Europe shares common political or governmental problems; (ii) common problems evoke common solutions; and (iii) common solutions sustain the progressive development of integrated politics. This practical, utilitarian way of looking at integration is, of course, a loose version of the old neo-functionalism. In practice, though not in logic, the idea of a common European *culture* tends to go hand in hand with the notion of common functional *problems*. Thus, the usually unthought-out[12] conception has been that 'European' government will come about to deal with 'European' problems for a self-consciously 'European' people sharing a 'European' culture. Where shared political problems are perceived, common culture underwrites common identity which incites communication and acceptance of common solutions; where there are common solutions, Europeans communicate more and further 'their' culture. Both these ideas suggest that one common thing leads more or less directly to another common thing, until a single, 'integral' entity emerges. It is that which lies at the heart of the unitarist thesis. So, I think of their combination as the impression that Europe is heading towards simple unity, i.e. an integration grounded in sharing 'the same', *singular* underpinnings (problems, culture, identity, etc.). It is the view that what Europeans have *in common* is shaping an integrated Europe. The 'unitarist thesis' maps out the integration of Europe in terms of this simple unity.

The first idea sustaining functionalist unitarism has been that government is a practical pursuit, intended to make social life work more smoothly and successfully for each and all. In order to function, that is to say, there is a need for organization that is shared right across society, and it is the business of politics and government to organize well. In the words of David Mitrany, functionalism: 'shift[s] the emphasis from political issues which divide to those social issues in which the interest of the people is plainly akin and collective'.[13] It is not difficult to think of aspects of the life of European societies which do indeed suggest common problems and hence a *need* for common solutions, administered by a common apparatus of government: trade extensive enough to permit economies of scale; the prevention of environmental damage across national borders; control of population movements; drug-trafficking; etc. The oddity is that many of the institutions and the policies apparently necessary to address the 'common' problems do not in fact materialize. Thus, from early on, neo-functionalist unitarism was criticized for its uncritical belief in the uncontentious spread of common solutions ('spillover' in the jargon) from one field of government to another (Caporaso, 1972).

The difficulty lies with the logic. From the fact that a society, or societies, may be thought to have one or more problems in common, it does not at all follow that the solution will be pursued in common, via common governance, or that it will be successful or popular. For a common problem is often not *perceived* as such. It may not be perceived at all, or be perceived as something else. A common problem may be seen as the *fault* of someone else who actually shares it. Thus the shortage of fish in the English Channel/Manche is seen not as a problem *shared* with the French/the English, but as the fault of the English/the French. And even if one group does realize that it has a problem in common with some other group, instrumentally speaking its best strategy may be to pursue a *separate* solution.[14] To take a view from the margins is to move away from all that. Whilst not *excluding* the possibility of things in common, it does not presuppose a unity in either the political challenges facing European societies or in their self-identities.

Accommodations with diversity

Whilst unitarism remains a natural *inclination* in accounting for integration, in the face of the hesitations and reversals of integration itself, integration theory has increasingly developed a scepticism about unitarism. The whole of the so-called 'inter-governmentalist' tendency in the analysis of integration can, for example, be seen to promote the view that integration is a matter of the concatenation of different (national) actors' pursuit of discrete interests which interlock into a merely apparent unity at the 'integrated' level. I want now to review that shift of perspective in the making. Whilst the marginal perspective of our discussion expresses the trend in the most explicit form, to scan the trend shows the extent to which the point of view we are piloting here can extend or deepen, rather than simply overturning, the responses of others.

Theoretical views which, in recent years, have given pride of place to the relations between sovereign governments, have a good deal of plain political realities backing them. In any European political process a prime element has always been MS governments' hope of gains in their own economic and political resources. Scratch the surface of a current EU issue – telecomms liberalization, open-skies agreements for air transport, environmental regulation, posted workers, etc. – and one finds MS governments assessing the chances of 'their' economies, 'their' nationalized industries, 'their' major companies, 'their' farmers or whatever gaining from it. In Andrew Moravcsik's 'liberal intergovernmentalist' model (1993), states make a positive assessment of their gains over their losses as integration evolves. They do this in a 'two-level game', played between MS governments at the upper level, and, at the lower, with a plurality of interests further afield, in domestic politics. Where intergovernmentalism entertains the idea, it defers unity to the outcome at the end of negotiations between diverse 'national' interests. 'Unitarism' is qualified, then, appearing only at a late stage (if at all). The character of the games in intergovernmentalism could, I suggest, be embraced under the account of actors on the margin: each MS (or sub-group of MSs) is able to declare itself *separate* from the others – either part of the majority or not – assessing the situation in terms of its own, separate position and exploiting its separateness. In other words, each MS acts as a margin *pro tem* in the sense I am exploring.

Intergovernmentalism has modified the role of the centre in unitarism, overshadowing the pull of central supranational institutions with the multipolarity of the long-standing European state system. It has found much prime material from historical studies, with their established (and inherently realist) taste for sovereign-state relations. Milward (1992), exploring the private calculations of the future MSs in the run-up to the treaties of the 1950s, is the *locus classicus* of the historian's scepticism towards unitarism. More recently, Robert Bideleux has examined the historical situation of Europe in the light of the break-up of the Communist Bloc (Bideleux and Taylor, 1996). He evokes an image of as much *dis*integration as integration: the 'eternal fluidity of European state systems', which has, moreover, been a base of Europe's *strength* in the past (p. 2f).[15] Faced with post-1989 restructuring, the European Union consists of 'quarrelsome, cantankerous and mutually competitive national states' under – Bideleux adds, evoking the two-level game of intergovernmentalist theory – 'politicians over-influenced by sectional producer lobbies' (pp. 243–4). On its own the historians' realism about European states altered the account of integration substantially. By formally incorporating the multipolar political environment *within* each sovereign state, Moravcsik's intergovernmentalism overlaid the state system of history with further pluralist domestic political orders. And Kersbergen (1997) has even produced a critique of EU and MS 'neo-voluntarist' retrenchment in social policy which moves between two-level intergovernmentalism and the redefinition of the boundaries of political entities. For Kersbergen argues that retrenchment is chipping away at national loyalty amongst European populations.

Intergovernmentalism can, on the other hand, easily be accommodated *alongside* the earlier unitarism: by setting up distinct institutional levels at which unitarist integration may be happening. Where that is done, certain institutions remain as sites for unitarism (classically the Commission, though the ECJ may also appear – Burley and Mattli, 1998). This development may even have been contained within neo-functionalism itself, insofar as it cast the Commission in the role of agenda-setter for the self-seeking, second-order debate among self-interested MSs (Lindberg, 1994). The 'new institutionalism', with which public-policy specialists have recently recovered some of the ground from intergovernmentalists, reasserted unitarism, located in the leadership towards unity undertaken by the institutional centre. Doing so, though, has required that the status of the centre be scaled down: Bulmer and Scott (1994) preferred to think a 'governance regime', something far less substantial than a state; while Majone (1993) emphasized the 'regulatory' character of the EU, so diluting the transnational uniformity of EU legislation on the grounds of its limited reach into society.

These debates between theories have gone a long way to replace naive unitarism, and hence the impression of a single, dominant centre for integration. But only recently, with the idea of 'networks' that generate policy outcomes at the European level, has a geometry for integration been proposed that accommodates diversity in its very form (Börzel, 1997). Networks are 'webs of relatively stable and ongoing relationships which mobilize and pool dispersed resources so that collective (or parallel) action can be orchestrated toward the solution of a common policy'.[16] The concept of a network is fundamentally non-unitarist, since networks can be built up in any order by making progressive links between 'nodes': no particular order of priority is implied and any given node in a network may link with any other. It has been particularly effective, then, in explaining why, in a field so diverse as EU policy formation, some players are able to enter and/or make a greater impact than others (Rhodes *et al.*, 1996). Abstractedly speaking, however, the network concept suggests rather flat structures, where the different nodes remain roughly on a par.

An implicit lacuna in the idea of a networks, is, then, the question of how those in a network possess different levels of power and tactical advantage. This is a difficulty which the marginal viewpoint addresses: since it includes the way that the parties on the supposed margins mark out, hold, create or claim positions of greater apparent centrality and power than others (Parker, 1998a). In incorporating these tactics, indeed, an account of integration from the margins crosses the path of constructivism, which has recently resurfaced as a view of how governmental entities come to be formed and re-formed (Jørgensen, 1997). Constructivism offers an ontology for the way that agents operate within given political 'realities' to invent new meanings and create new structures: the account of marginality that I have explored here shares, in respect of the margins, constructivism's interest in how agents can create and re-create political givens. In examining cases of the politics of marginality, the third part of this book in particular illustrates strategies whereby players on the margins may, for example, seek to claim central position, or shift the meaning of the centre in their favour.

I have tried in this chapter to define what margins are and to establish why they may be important for integrated Europe. I suggest that focusing upon the margins offers a common way forward to see the impact of diversity in both contemporary European integration and the long-term expansions and contractions of European 'civilization' (Parker, 1998b). The advantage of developing a common way for both areas of inquiry is that each is made stronger. With EU enlargement very much in view, we should bear in mind that the long history of Europe's expansion, contraction and re-expansion is an ongoing story. The discussion of this book overall may be seen as a first attempt to apply a margins research agenda to that story. The chapters that follow consider many particular economic, social or cultural areas where what is done at the margins may be a force for structural change in Europe as a whole. They are samples of what may be learnt by refreshing the research agenda with an emphasis on what would otherwise be thought of as 'merely' marginal.

Notes

1 See also Judt (1996), which is further considered in Chapter 11, for an argument to the effect that the disappearance of the Cold War border has knocked away the rickety ground of a common self-identity supposed to sustain European integration.
2 Furthermore, the common root word for the two appears to be the old Norse *mörk*, in turn the source of English and Scandinavian words for darkness.
3 Hirschman's analysis embraces the customer–supplier relationship as well as the organization–member one – indeed, the former is his starting point. Nonetheless, I ignore that for present purposes, assuming that it would be jejune to insert '*mutatis mutandis* repeatedly for the customer–supplier relationship'.
4 A significant instance of something like that has been theorized by Sperling and Kirchner (1997), who try to model the difficulties of constructing a security architecture for a Europe where the 'other' side becomes your own side and military measures of 'security' fuse with economic ones. The boundaries of the entity defended and of security policy (as against other policies) have both collapsed, and the logic of defence becomes accordingly difficult to re-establish.
5 Cf. Anderson's account of the range of political choices invoked by migration across national frontiers (1996, pp. 127ff).
6 E.g. Norval (1996) and Tin (1998) offer suggestive accounts of how Afrikaaners wrestled – unsuccessfully – with a marginal 'European' identity on their rejected edge of European colonial space.
7 '... the time has come to combine [de Gaulle's concept of the nation] with a requirement for "integration" inside and outside its borders' (p. 68).
8 '... this undecidability is what I call politics... wherever there is a frontier, be it between two countries, be it between literature and philosophy..., there is something of the order of politics' (Bennington, 1996, p. 3). This style of thought is also explored in the context of sustaining national identity in Bhabha, 1990.

9 These in turn are derived from the move, led by Holsti and Rosenau in the 1970s, away from the hard-edged realism of an international relations theory grounded in power and national interest and towards analysis in terms of societal or other values.

10 Elsewhere in Mouritzen *et al.*, Håkan Wiberg (pp. 43–63) does indeed suggest other categories ('aspiring', 'resigned' or 'voluntary' outsiders). Likewise, Mouritzen himself (p. 12) envisages both passive adaptations to the environment constellation and more assertive postures of 'dominance' or 'balance'. These postures do not, however, figure in the final results for the countries looked at.

11 It may be relevant to observe that the very expression 'integration' conveys an implication of this kind. For it contains a reference to the wholeness, 'integrity' or 'integral' existence, of the entity that will *result* from the process: Europe itself. Hence its etymological link to the latin *integere,* to protect undamaged or preserve as whole.

12 Karl Deutsch's approach (Deutsch and Burrell, 1957), is a behaviourally sophisticated version of this kind of unitarism, which has consequently survived as a theoretical classic. It emphasizes communication and culture in the sense of common *understanding* between populaces and political elites in different component parts forming an integrated entity. Initially, there is not so much common understanding as 'complementarity, that is, an interlocking relationship of mutual resources and needs' (p. 90). But Deutsch's unitarism emerges in his expectation that this represents merely a *transitional* state of affairs, which would normally be superseded in due course by a more solid, unitarist community. It is susceptible to a counter-argument analogous to that set out below against unitarism in general: namely, culture and communication links across an integrated population do not necessarily have to be interpreted as unitary (Zetterholm, 1994).

13 *International Affairs*, XXIV (1948), p. 356, quoted in Haas (1964), p. 7.

14 This is a version of the 'free-rider' problem in the theory of rational preferences. Where a number of groups or individuals are in a situation together, there is always the possibility that one party may see a way to enjoy a benefit/avoid a loss without contributing to the cost of the solution.

15 Gerschenkron (1962) once evoked the idea that differing degrees of industrialization allowed a leapfrogging development by the backward which lay behind Europe's overall productive energy.

16 P. Kenis and V. Schneider (1991) 'Policy Networks and Policy Analysis: Scrutinizing a New Analytical Toolbox', in B. Marin and R. Mayntz (eds), *Policy Network: Empirical Evidence and Theoretical Considerations* (Frankfurt-am-Main: Campus Verlag), p. 36; quoted in Börzel, 1997, pp. 12–13.

References

Albrow, M. (1996) *The Global Age: State and Society Beyond Modernity* (Cambridge: Polity).

Anderson, M. (1996) *Frontiers: Territory and State Formation in the Modern World* (Cambridge: Polity).
Bartlett, R. (1993) *The Making of Europe: Conquest, Colonisation and Cultural Change, 950–1350* (London: Allen Lane; Penguin Press).
Bennington, G. (1989–92) 'Four Lectures about Frontiers'. University of Sussex:http://members.aol.com/gbennin785/frontier.htm.
Bennington, G. (1996) 'Frontiers of Literature and Philosophy'. Inaugural lecture (Brighton: University of Sussex).
Bhabha, H. (1990) 'Dissemination: Time, Narrative, and the Margins of the Modern Nation', in H. Bhabha (ed.), *Nation and Narration* (London: Routledge), pp. 291–322.
Bideleux, R. and Taylor, R. (eds) (1996) *European Integration and Disintegration: East and West* (London: Routledge).
Bleiker, R. (1997) *Retracing and Redrawing the Boundaries of Events: Postmodern Interferences with International Theory*, DUPI Working Paper no. 1997/17 (Copenhagen: Dansk Udenrigspolitisk Institut).
Börzel, T. J. (1997) *Policy Networks: A New Paradigm for European Governance?* EUI Working Paper no. RSC 97/17 (Florence: European University Institute).
Breuilly, J. (1998) 'Sovereignty, Citizenship and Nationality: Reflections on the Case of Germany', in M. Anderson and E. Bort (eds), *The Frontiers of Europe* (London: Pinter), pp. 36–67.
Bulmer, S. and Scott, A. (eds) (1994) *Economic and Political Integration in Europe: Internal Dynamics and Global Context* (Oxford: Blackwell).
Burley, A.-M. and Mattli, W. (1998) 'Europe Before the Court: a Political Theory of Legal Integration', in B. F. Nelsen and A. C.-G. Stubb (eds), *The European Union: Readings on the Theory and Practice of European Integration* (London: Macmillan), pp. 241–71.
Caporaso, J. A. (1972) *Functionalism and Regional Integration: A Logical and Empirical Assessment* (Beverly Hills, CA).
Caporaso, J. A. (1996) 'The European Union and Forms of State: Westphalia, Regulatory, or Post-Modern?', *Journal of Common Market Studies*, **34**(1), 29–52.
Christiansen, T. (1997) 'Reconstructing Space: from Territorial Politics to European Multilevel Governance', in K.-E. Jørgensen (ed.), *Reflective Approaches to European Governance* (London: Macmillan), pp. 51–68.
Clark, J. (1991) 'Britain as a Composite State – Sovereignty and European Integration', in U. Østergård (ed.), *Britain: Nation, State and Decline* (Copenhagen: Academic Press), pp. 55–83
Derrida, J. (1992) *The Other Heading: Reflections on Today's Europe*, trans. P.-A. Brault and M. B. Naas (Bloomington, IN: Indiana University Press).
Deutsch, K. W. and Burrell, S. (eds) (1957) *Political Community and the North Atlantic Area: International Organization in the Light of Historical Experience* (Princeton, NJ: Princeton University Press).
Duff, A. (1997) *Reforming the European Union* (London: Federal Trust/Sweet & Maxwell).
Gamble, A. (1991) 'From Empire to De-industrialisation', in U. Østergård (ed.), *Britain: Nation, State and Decline* (Copenhagen: Academic Press), pp. 85–103.
Gerschenkron, A. (1962) *Economic Backwardness in Historical Perspective* (Cambridge, MA: MIT Press).

Guéhenno, J.-M. (1995) *The End of the Nation-State*, trans. Victoria Elliott (Minneapolis, MN and London: University of Minneapolis Press).
Haas, E. B. (1964) *Beyond the Nation-State: Functionalism and the International Order* (Stanford, CA: Stanford University Press).
Hirschman, A. O. (1970) *Exit, Voice and Loyalty: Responses to Decline in Firms, Organizations, and States* (Cambridge, MA: Harvard University Press).
Jørgensen, K. E. (ed.) (1997) *Reflective Approaches to European Governance* (London: Macmillan).
Judt, T. (1996) *A Grand Illusion: An Essay on Europe* (London: Penguin).
Kersbergen, K. van (1997) *Double Allegiance in European Integration: Publics, Nation-States and Social Policy*. EUI Working Paper no. RSC 97/15 (Florence: European University Institute).
Kristeva, J. (1993) *Nations without Nationalism* (New York: Columbia University Press).
Lenoble, J. and Nicole, D. (1992) *L'Europe au Soir du Siècle: Identité et Démocratie* (Paris: Editions Esprit).
Lindberg, L. (1994) 'Comment on Moravcsik', in S. Bulmer and A. Scott (eds), *Economic and Political Integration in Europe: Internal Dynamics and Global Context* (Oxford: Blackwell), pp. 81–5.
Majone, G. (1993) 'The European Community between Social Policy and Social Regulation', *Journal of Common Market Studies*, **31**(2), 153–68.
Marks, G., Scharpf, F. and Schmitter, P. (1996) *Governance in the European Union* (London: Sage).
Milward, A. (1992) *The European Rescue of the Nation-State* (London: Routledge).
Minghi, J. V. (1991) 'From Conflict to Harmony in Border Landscapes', in D. Rumley and J. V. Minghi (eds), *The Geography of Border Landscapes* (London: Routledge), pp. 15–30.
Mouritzen, H. (1998) *Theory and Reality of International Politics* (Aldershot: Ashgate).
Mouritzen, H., Wæver, O. and Wiberg, H. (1996) *European Integration and National Adaptation: A Theoretical Inquiry* (New York: Nova Science Publishers).
Moravcsik, A. (1993) 'Preferences and Power in the European Community: a Liberal Intergovernmentalist Approach', *Journal of Common Market Studies*, **31**(1), 473–523.
Norval, A. (1996) 'Social Ambiguity and the Crisis of Apartheid', in E. Laclau (ed.), *The Making of Political Identities* (London: Verso), pp. 115–37.
O'Dowd, L. and Wilson, T. M. (1996) *Borders, Nations and States* (Aldershot: Avebury).
O'Sullivan, N. (1993) 'Political Integration, the Limited State, and the Philosophy of Postmodernism', *Political Studies*, **XLI** (Special Issue – 'The End of "isms"'), 21–42.
Parker, N. (1998a) *Geometries of Integration*. DUPI Working Paper no. 1998/15 (Copenhagen: Dansk Udenrigspolitisk Institut).
Parker, N. (1998b) *The Ins-and-outs of European Civilization: How Can the Margins Inform Research on Europe?* CFK Working Paper no. 69–98 (Århus: Center for Kulturforskning, University of Århus).
Rhodes, R. (1996) 'Towards a Postmodern Public Administration: Epoch, Epistemology or Narrative?', *Newcastle Discussion Papers in Politics*, vol. 16 (Newcastle: University of Newcastle).

Rhodes, R., Bache, I. and George, S. (1996) 'Policy Networks and Policy-making in the European Union: a Critical Appraisal', in L. Hooghe (ed.), *Cohesion Policy and European Integration: Building Multi-Level Governance* (Oxford: Oxford University Press), pp. 367–87.
Risse, T., Engelmann-Martin, D., Knopf, H.-J. and Roscher, K. (1998) *To Euro or Not to Euro? The EMU and Identity Politics in the European Union*. EUI Working Paper no. RSC 98/9 (Florence: European University Institute).
Rosecrance, R. (1986) *The Rise of the Trading State: Commerce and Conquest in the Modern World* (New York: Basic Books).
Selm, B. van (1997) *The Economics of the Soviet Break-up* (London: Routledge).
Sperling, J. and Kirchner, E. (1997) *Recasting the European Order: Security Architectures and Economic Cooperation* (Manchester: Manchester University Press).
Story, J. (1997) 'The Idea of the Core: the Dialectics of History and Space', in G. Edwards and A. Pijpers (eds), *The Politics of European Treaty Reform: The 1996 Intergovernmental Conference and Beyond* (London: Pinter), pp. 15–43.
Stubb, Alexander C. G. (forthcoming), *Flexible Integration and the Amsterdam Treaty: The Dynamics of EU Negotiations* (Basingstoke: Macmillan).
Tilly, C. (1992) *Coercion, Capital and European States, AD 990–1992* (Oxford: Blackwell).
Tilly, C. (1995) *Popular Contention in Great Britain, 1758–1834* (Cambridge, MA: Harvard University Press).
Tin, H. (1998) *The Slave, the Native and the Serf: Three Sites of Violence in the South African Ethnic Spaces 1652–1994*. CFK Working Paper, no. 59–98 (Aarhus: Center for Kulturforskning, University of Aarhus).
Tsoukalis, L. (1996) 'Economic and Monetary Union', in H. Wallace and W. Wallace (eds), *Policy-Making in the European Union* (Oxford: Oxford University Press), pp. 279–99.
Weber, E. (1976) *Peasants into Frenchmen* (Stanford, CA: Stanford University Press).
Zetterholm, S. (1994) 'Why is Cultural Diversity a Political Problem? A Discussion of Cultural Barriers to Political Integration', in S. Zetterholm (ed.), *National Cultures and European Integration* (Oxford: Berg), pp. 65–82.

2

European Integration through the Kaleidoscope: the View from the Central and East European Margins

Helen E. Hartnell

Locating Europe's margins, or the margins of Europe

The notion of 'Europe's margins' suggests that there is a European entity – if not a centre – from which the edges can be perceived. In the context of integration studies this notion appears to coincide with the unilateral view of European integration as a matter of European Union (EU) enlargement, largely as a structural problem of integrating developing into developed countries. The goal here is not to devalue this perspective, but to expose its EU-centric bias and put it in its proper place (Wilson and van der Dussen, 1995). The disordered post-1989 'New Europe' looks quite different from the post-war bipolar order engraved on our minds (Budge *et al.*, 1997; Hadas and Vörös, 1997; Judt, 1997; Pocock, 1997; Agh, 1998). A decentred vision of Europe must bring into focus the area peripheral to integrative entities or centres such as the strong Brussels-based EU and the increasingly weak Moscow-based Commonwealth of Independent States (CIS) (Shinkarenko, 1996). Peripheral areas such as EFTA and the EEA, Central Europe, the Balkans, and the Baltic States might be seen as located on the margin of, or even in between, the EU and the CIS. It is in this geographical sense that I use the term 'margin'. This decentred vision must be contrasted with the view of Europe from the EU's own perspective, from which there is a series of concentric circles, of ever more remote fringes or margins, with Russia and the CIS occupying the outermost position. This vision is implied in the EU's use of the term 'Central and Eastern Europe' to include all non-EFTA and non-EEA European countries.

This chapter visits multiple non-EU perspectives on European integration, and concludes that the integration process has become not just a multipolar one, but a manifold one, in which initiative and innovation can be found in those remote, off-centre, in-between, marginal sites. This conclusion follows from an examination of the phenomenon of subregional coalescence in Central and Eastern Europe (Kolankiewicz, 1994, p. 6; Hartnell, 1997). Subregional coalescence calls for 'a greater tolerance for those that are small, that do not cast a long shadow in the world of today' and that may 'be the engines of change, both by the disorders that they produce but also by the ways in which their values, however seemingly antiquated, survive and prosper, and not only in their native lands' (*Daedalus*, 1997, p. xi). This solidification of the margins suggests the advent of a counterweight to the main European regional integration initiatives – *viz.* the EU and the CIS – and evidences some ways in which the post-1989 European integration process is decentred and transnational, if not entirely postcolonial (Rushdie, 1996, p. 50).

This chapter also examines the dynamic relationship between the EU and certain Central and East European non-member countries that are attracted by its gravitational pull (Peers, 1995; Mayhew, 1998). This analysis both highlights problems with the EU's top-down, vertical integration strategy *vis-à-vis* its neighbours, and shows how subregional coalescence can tilt the integration process on its fulcrum, towards the horizontal, and thereby ameliorate the pathologies inherent in the EU's vertical strategy. Thus, in the end, the analysis concerns the salutary prospect of intermediation by marginal actors, by means of participation in various new formal and informal institutions.

Viewed in this way, marginality means more than simply being outside of or peripheral to a whole, more than making the best of a bad situation. Marginality implies a site where bold innovation and nimble action are possible, a site at which Central and East European non-member states must act to advance their own interests, and from which they can try to move Europe. I would, with Parker (see p. 8), 'dissociate "marginality" from the idea of dependence upon, or inferiority to, a corresponding "core"'. Marginal actors, in this sense, are wheels that turn, and not simply cogs that are turned by greater forces. Margins, therefore, constitute one 'source of dynamism [to] take the place of [Europe's] historic antagonisms' (Malia, 1997, p. 20).

Parker's broader emphasis on margins as an alternative to unitarist theories of integration is both timely and welcome. Subregional coalescence evidences Parker's hypothesized 'many' intersecting processes of integration, his 'shifting, overlapping and interlocking entities'. Yet this notion represents more than just a new word for intergovernmentalism (Moravcsik, 1993), for liberal bargaining writ large: it entails embracing the notion of European integration as a decentred process, and necessitates stepping away from an EU-centric focus and valorizing the diverse coalitions that emerge in various locales.

This conception of integration goes beyond 'multi-speed' Europe, beyond even Parker's vision of 'concatenation' and the EU's perpetual accommodation between margins. I not only look at how the EU accommodates margins, but advocate a move towards a truly marginal perspective on the integration process. This chapter is partly dedicated to considering the ways in which Central Europe can meaningfully become a centre, albeit positioned between the two main regional integration engines.

Integration is thus not one process, but many, and there is not just one centre of Europe, but many. These various integration projects cannot be viewed in isolation from one another. They occasionally complement or conflict with, often overlap with, and frequently influence one another. A multipolar vision of Europe that includes marginal perspectives requires that we abandon the EU's telescopic, distancing vision of integration, in favour of a kaleidoscopic one, which allows us to see Europe's full array of parts and patterns, continually rearranging themselves in new and intriguing ways.

The view from the margins: the EU's association strategy

Seen from the countries of Central and Eastern Europe (CEECs), the EU's integration strategy appears to be of a vertical, top-down nature. The EU has prepared for its neighbours various types of bilateral treaty to establish the 'trade-plus' relationship technically known as 'association', and has in fact concluded such agreements with most (but not quite yet all) CEECs. The typologies and features of these bilateral treaties are examined in depth elsewhere, as is the status of the EU's relationship with each of the CEECs (Hartnell, 1997, pp. 119–51; 1993a and 1993b). Here only a few structural features will be described, before moving to an examination of the tensions in the relationship between the EU and the CEECs. These tensions illustrate the pathologies of the vertical association relationship, and thereby set the stage for analysing the effects of subregional coalescence.

One further general feature of the EU's association strategy bears preliminary mention: its tendency to utilize both bilateral and multilateral approaches to the CEECs. The EU has negotiated separate bilateral treaties with nearly all countries in Central and Eastern Europe ('hub-and-spoke' structure), yet often treats the associated CEECs as a bloc and encourages them to develop relationships among themselves on a subregional basis. This complex strategy often vexes the associated countries, which insist on being treated as autonomous, sovereign states, and resent being forced back into a collective.

Key features of association

The EU's association strategy *vis-à-vis* the CEECs is highly differentiated. Despite many shared features there is a fundamental distinction between those association agreements which are designed to lay the foundation for full political and economic integration of the associated country into the EU (so-called Europe Agreements, or EAs), on the one hand, and those which are not (so-called Partnership and Cooperation Agreements, or PCAs), on the other. The EU has concluded a 'pre-accession' EA-type of association treaty with Bulgaria, the Czech Republic, Estonia, Hungary, Latvia, Lithuania, Poland, Romania, Slovakia, and Slovenia. Each of these EA countries has formally applied to join the EU, thus confirming the failure of the EEA (Dupont, 1998). Other CEECs are relegated to the PCA circle, and have little prospect of being candidates for EU membership any time soon.

One of the key *de-jure* distinctions between EAs and PCAs is that the former purport to establish free trade areas (at least with respect to certain goods), and thus involve inherently preferential trading arrangements, whereas the latter involve trade normalization without commitment to further liberalization. Another significant *de-jure* distinction is that EAs require each associated country to 'approximate' or harmonize its legislation to that of the EU, whereas PCAs contain no such prescription. The EA countries must essentially remake themselves in the EU's image if they want to be eligible to join. Finally, there is one *de-jure* similarity which masks a *de-facto* distinction between the EA and PCA types of association relationship. While apparently similar in that both types of treaty call for political dialogue and create weak institutional structures, the EU has in fact developed a host of structures which apply to EA but not to PCA countries, such as the structured dialogue.

The pathologies of association

The EU's framework for wider European integration is beset with problems. Elsewhere I have characterized these problems as the five paradoxes of association (Hartnell, 1997), since the contradictions appeared to be intractable. Here, however, I reconceptualize some of the problems as pathologies that might be susceptible to cure. In any case, two types of problem must be distinguished: the first pertaining to imbalances or contradictions that inhere in the 'vertical' relationship between the EU and the associated CEEC countries, and the second to tensions that inhere in the 'horizontal' relationship among the associated CEECs themselves.

In both types of case the problems cannot be blamed solely on any one party, though it is easy to condemn the EU hegemon for its neocolonial posture towards its neighbours. And yet one cannot lay the whole blame on the EU, since some pathologies have resulted from the CEECs' own conduct. Many CEECs undertook hasty unilateral trade liberalization and engaged in 'anticipatory adaptation' (i.e. unilateral voluntary convergence to EU norms), and thus weakened their bargaining position by 'making concessions prior to securing any *quid pro quo*' (Haggard *et al.*, 1993, p. 188). Bárta and Richter (1996, p. 24) later confirmed that the EAs establishing free trade areas, despite their many advantages 'paradoxically . . . seriously weakened the [EA countries'] bargaining position, because the main and perhaps only relevant asset that [they] were able to offer was free access for EU exporters to domestic CEEC markets'. Still, it is difficult to blame the CEECs for having harboured unrealistic expectations of the EU, or for gullibly sacrificing long-term interests for short-term gains, especially considering the urgency of the period immediately after 1989. This short-term mentality has faded as CEECs have realized that accession will neither be quick, nor provide a panacea for existing problems.

The carrot of EU membership is less tempting today than in the aftermath of 1989, for two reasons. First, the carrot looks smaller, because it recedes into the distance as the stick dangling it grows ever longer. Second, the carrot is sure to shrink in real terms upon reform of the EU's redistribution schemes (*viz.* Common Agricultural Policy and the Structural Funds). It is no wonder that the motivational force created by this arrangement has become attenuated, and that the Central/East European donkey – hitched up to pull the cart of economic growth – is getting some ideas of its own. CEEC (and in particular EA) countries have been showing signs of stubbornness and even reorientation. They recognize that the current vertical East–West European integration model is imbalanced in the EU's favour, but have discovered that they are not powerless to change (or at least influence) the direction of movement. Analysis of five problems that beset the EU's current vertical integration model is a prerequisite to both an appreciation of the recent changes in Central and Eastern Europe, and to an understanding of the forces working for and against European integration.

Trade

CEEC countries suffer from a high level of foreign debt, and are keen to earn foreign exchange. The EU, on the other hand, has constructed an intricate network of external trade treaties, including but not limited to those with CEEC countries, which are designed in part to realize the goal, stated in the Commission's 1996 Market Access Strategy, to use free trade agreements as tools to promote market access for EU exporters. This is equally, if not especially, applicable in the case of CEEC countries.

The free trade provisions in the EA association agreements illustrate this partially incompatible set of urges. At first the EAs were touted for the asymmetric nature of their transitional provisions. On paper the EAs require the EU to open up its market to imports from each EA country faster than the associated country must open up to imports from the EU. This paper sacrifice, however, masked a structural imbalance. The EA regime inherently favours the EU, since it provides for unlimited trade of industrial goods, which the EU tends to export, but restricts trade in sensitive goods, such as agricultural products, textiles, coal and steel, which EA countries wish to export to the EU. Implementation of the EAs tended to be accompanied by a dramatic rise in EU exports to the EA countries, without an offsetting rise in EU imports from Central and Eastern Europe. Thus, for example, the first six EA countries (Bulgaria, the Czech Republic, Hungary, Poland, Romania and Slovakia) went from being net exporters of agricultural products to net importers in 1994 and 1995 (Bárta and Richter, 1996, p. 1). This structural imbalance continues to spur accusations that the EU is abusing its dominant economic position.

Aspirations

The contradictions inherent in the aspirations of the EU and EA countries exist at two different levels. First, the question is whether the desire for East–West union is truly a shared goal. Second, even if that goal is a shared one, there are incongruent notions of when and why union should occur. Today there is formal agreement – at least at the governmental level – that East–West union is a shared goal. There is also a working assumption that some EA countries will indeed become members of the EU after the year 2000. Yet these apparent congruities mask deep divisions over the prospects for EU enlargement and many perceive the EU as trying to slip the noose of its commitment to enlargement. The discrepancy, which first manifested itself as a disagreement over timing, reflects a fundamental difference of opinion as to the rationale for enlargement.

The EU thinks of membership as a reward for successful transition, whereas the EA countries view membership as the key to successful transition. In other words, the EU says that the EA countries may join when they are ready to be full members, while the EA countries consider full membership necessary in order to achieve that readiness. In Nikolaïdis' words, 'was regulatory change within eastern European countries seen as a precondition or an effect of closer ties' with the EC (1993, p. 236)? This dilemma manifests what Lord Dahrendorf called the 'incompatible time-scales of political and economic reform' and epitomizes the true distance between the EU and its neighbours in Central and Eastern Europe. The EA countries pushed consistently during the early 1990s for a firm commitment from the EU as to when accession negotiations would begin. They managed to extract a series of promises that talks would begin within a certain period of time after conclusion of the Intergovernmental Conference (IGC). These commitments turned out to be moving targets, in part because the IGC was not concluded on schedule, and in part because the EU repeatedly postponed the commencement of enlargement negotiations.

In July 1997 the Commission presented its long-awaited 'Agenda 2000' report and opinions (*avis*) assessing the readiness of each applicant country. The Commission recommended that accession negotiations be opened with Cyprus, the Czech Republic, Estonia, Hungary, Poland, and Slovenia – but not with Bulgaria, Latvia, Lithuania, Romania, Slovakia or Turkey. The Luxembourg European Council, meeting in December 1997, followed suit and decided to convene bilateral intergovernmental conferences to begin negotiations with the six 'fast-track' countries. In an attempt to soften the blow to the second tier of countries, the EU decided formally to commence enlargement for all applicant countries at the first meeting of the European Conference in London on 12 March 1998, but actually to start negotiations with the six fast-track countries in Brussels on 31 March 1998. Yet the fanfare of early 1998 was followed by further delays, and 'genuine accession negotiations at ministerial level' began only in November 1998 (*Agence Europe*, 14 October 1998, p. 8).

This sequence of events predictably upset the second tier of countries and left them in grave doubt as to the 'shared goal' of eventual membership. Commissioner van den Broek sought to reassure them that the 'natural differentiation among the applicants for a variety of historical, political and economic reasons . . . in no sense means discrimination' (*Agence Europe*, 17 July 1994, p. 4). He further stated that 'all applicants are assured of membership' and that 'it is not a question of if, but when' (*Agence Europe*, 6 June 1997, p. 9). The second-tier EA countries were told that they 'can move to the "fast-track" for prospective new members if they make sufficient progress in economic and political reform' (*RFE/RL Newsline*, 15 December 1997), but for a long time no criteria or procedures were devised to determine how or when to adjust their status.

The dispute over commencement of the enlargement and accession negotiations has now been formally resolved, at least for those countries on the fast track, though procedural hairs continue to be split. For the most part, attention has shifted onto substantive issues, where serious obstacles remain. On the EU side it is clear that the IGC failed to undertake the institutional reforms and policy changes that are prerequisites to the next wave of enlargement. In mid-1997, when accession negotiations would have commenced if the EU had kept its earlier promises to the CEEC countries, the Commission confessed that 'before facing the complexities of enlargement, [the EU] needs to address internal problems as well as come up with a strategy for the future which includes the uncertainties of enlarging the union' (*European Dialogue*, July/August 1997). So there will probably have to be yet another IGC prior to actual accession by any of the applicant countries, and countless further opportunities to scrutinize the accession criteria stated by the Copenhagen European Council in June 1993, which include the EU's 'capacity to absorb new members, while maintaining the momentum of

European integration'. The enlargement process has been infected by 'unease, and even a certain mistrust' (*Agence Europe*, 5 and 6 October 1998, p. 3).

On the CEEC country side, the obstacles to accession reside mainly in the remaining accession criteria stated in Copenhagen: stability of institutions guaranteeing democracy, the rule of law, human rights, and respect for the protection of minorities; a functioning market economy as well as the capacity to cope with competitive pressure and market forces within the EU; and the applicant's ability to take on the obligations of membership – the *acquis communautaire* – and adhere to the aims of political, economic and monetary union. Not even the most advanced EA countries will have an easy time fulfilling all of the quantitative economic and monetary conditions for accession, and even such a country will encounter further difficulties in keeping up with the expanding *acquis* and in satisfying the amorphous political criteria, all of which pose qualitative barriers to full membership. Indeed the *acquis* is expanding precisely in areas, such as the Third Pillar, which will erect high hurdles to CEEC membership. In addition, the EU has made it increasingly clear that it is not enough for the applicant countries to approximate their laws to those of the EU; they must also implement those laws in a satisfactory manner, and so the EU has begun to emphasize reform of public administration and the judicial system. The Commission has stated (*European Dialogue*, July/August 1997):

> Imagine being asked to build a state-of-the-art spacecraft using nothing more than a cheap screwdriver and a handful of nails. This is the scale of the task facing many of the [CEEC applicant countries which] need to put in place an efficient public administration system capable of coping with EU legislation. . . . All applicants have theoretical programmes in place for taking on the *acquis*, but that is not necessarily the point. It is relatively easy to adopt legislation in national parliaments. Making it function on the ground is another matter.

Confronted with such an official attitude, one can hardly refrain from doubting the sincerity of the EU's declared goal. But at the same time, it is readily apparent that the EU is devoting considerable resources (via PHARE and other funding programmes) to helping the applicant countries to satisfy the accession criteria.

Participation

The most disturbing (and intractable) pathology in the context of EU–CEEC relations concerns the gap between democratic rhetoric and reality. It is undisputed that the EAs are meant to encourage the transition towards democratic government as well as market economies in Central and Eastern Europe. And yet the EU's vertical association strategy is profoundly anti-democratic, as Ann Kennard (Chapter 7) points out in the case of Poland, for it has denied the EA countries an opportunity to participate in decisions or even discussions which affect their interests. The need for meaningful participation rights at the international (i.e. EU) level is especially important, since the 'internationalization of democratic rhetoric has accompanied a domestic displacement of democratic politics' (Kennedy, 1991, p. 384).

The deficiencies in the EU's strategy are of both substantive and procedural character. The EA countries have been denied participatory rights in two substantive areas: approximation of laws and the IGC, whereas the procedural deficiencies inhere in the Structured Dialogue itself.

The sweeping requirement that EA countries approximate their laws to those of the EU does not carry with it any rights to participate in the process of making or modifying EC law. This obligation, while a common element of the EU's accession policy towards all prospective members, is particularly offensive when imposed on Central and Eastern European countries, where transition and enlargement are intimately tied to the promotion of democracy. With regard to the IGC, the EA countries repeatedly requested (and were denied) the opportunity to participate. While the IGC was surely concerned with affairs internal to the EU, it is equally true that excluding the CEECs (and other applicants for membership) from participation was tantamount to excluding them from the drawing board for the 'new' Europe. Unlike some tensions which abate over time, this one grows ever stronger.

Procedural flaws afflict the Structured Dialogue (established pursuant to the 1993 Copenhagen European Council), which constitutes the principal framework for ongoing participation between the EU and the EA countries. The Structured Dialogue is a framework for regular political cooperation – on a multilateral basis – between the CEECs and EU representatives. The EA (and particularly the Visegrád Group) countries have viewed this format sceptically since its inception, not least because of its multilateral nature. Criticism and calls for improvement have been continual. The most consistent complaint has been that meetings are often 'ceremonial' in nature and do not provide adequate opportunity for discussion.

Sovereignty

The perennial issue of sovereignty plays an important role in the newly emerging democracies in Central and Eastern Europe, as it continues to do within the EU itself. While related to the former discussion of democratic participation, the sovereignty problem warrants special, if brief, mention. Each EA country experiences the contradictory impulses to revel in newly won sovereignty, on the one hand, and to join the EU and thereby surrender that sovereignty, on the other. Many people in the region share the belief that 'independence must precede interdependence' (Bielski, 1995, p. 17). Polish Primate Józef Glemp noted that membership 'is not just a political issue. The Church perceives it as a moral problem as well . . . a fight for independence, for preserving one's identity.' Reasoned debates over the nature of the EU's risk to national sovereignty aside, this sentiment should not be ignored, lest it 'sour' into 'anti-European feeling' or (worse) fester into nationalism and xenophobia (Kolankiewicz, 1994, p. 482).

The sovereignty pathology will continue to afflict relations between the EU and the applicant countries, particularly as citizens of EA countries learn that accession entails costs as well as privileges. Overall, however, this problem should abate, since the discourse is likely to shift away from traditional concerns with sovereignty, and towards modern concerns about national identity.

Competition or cooperation?

The final and most intriguing feature of the current integration model is a paradox which afflicts the horizontal relationship of the CEEC (and in particular the EA) countries amongst themselves. These countries are torn between the need and desire to cooperate and to compete with one another.

There are many reasons for the EA (and PCA) countries in Central and Eastern Europe to cooperate with each other. The CEEC countries – notwithstanding their differences – have a great deal in common, including their shared history, the difficulties of their current transition processes, and (for some) their fixation on the EU as the road to salvation. The EU has urged the CEEC countries to cooperate with each other, in part because it rightly knew it could not fill the void left by the post-1989 collapse of COMECON trade, and in part because of the potential benefits to be gained through shared transition experiences. 'What each country lacks individually could be compensated for collectively' (Kolankiewicz, 1994, p. 480). But these countries have often been reluctant to cooperate with one another.

At first the CEECs – and especially the Visegrád countries, *viz.* Hungary, Poland, and (then) Czechoslovakia – resisted being forced into anything resembling the old COMECON bloc. Indeed, as Nikolaïdis notes, they decided to 'scrap past patterns of cooperation' (1993, p. 201) in January 1990, and presided hastily over COMECON's demise (p. 240), not least since they viewed the 'trade patterns created by [COMECON] as artificial deviations from those that would have been created by their "natural" comparative advantage' (p. 201). Moreover, the Visegrád group feared being forced to proceed on the road to accession at the speed of the slower EA countries (e.g. Bulgaria and Romania), or worse, being consigned permanently to Europe's periphery. They were reluctant to associate with one another, lest that cooperation seal their marginal fate.

These attitudes persist, in some measure, but have been largely supplanted by a growing willingness to cooperate for mutual advantage. Indeed, many EA countries are gradually realizing the intermediary power that exists at the margins, in their 'unique corridor' (Lengyel, 1995) space in between Brussels and Moscow. Unlike Lengyel, who laments the lack of 'clear boundaries among the various [integration] models', I see great opportunity in this fluidity, and welcome the subregional coalescence described below. Yet these opportunities can be realized only to the extent the CEEC countries overcome their reluctance to cooperate, and avoid destructive competition with one another.

While Hungarian and Polish leaders claim that their countries 'are partners, not rivals' in their bid for EU membership, it is undeniable that the competition for EU entry has 'served to breed mutual suspicion and intimations of national superiority among the contestants' (Kolankiewicz, 1994, p. 482). EA countries have often vied to outdo each other in jumping through the pre-accession hoops held up by the EU. Joint ministerial meetings, held within the Structured Dialogue framework, offer a major arena in which EA countries can manoeuvre for advantage over one another, and where the EU can play them off against each other.

The final feature of the cooperation–competition paradox is more perverse, though less harmful, than the previously noted one. The applicant countries, while deeply suspicious of the EU's motives and ambivalent towards each other, have discovered that cooperation among themselves provides yet another opportunity for them to compete with each other for the EU's approval. Whatever impels it, cooperation promises significant benefits to the CEEC countries which embrace it. This chapter now turns to an examination of the phenomenon of subregional coalescence, and later, to its effects on the European integration process.

Subregional coalescence in Central and Eastern Europe

New regional and subregional initiatives have proliferated in Europe (as elsewhere) in recent years. The growing importance and self-confidence of subregional integration in Central and Eastern Europe introduces a new dynamic into the process of European integration. Subregional coalescence leads to migration of the centre and redefinition of boundaries, and offers the vision of kaleidoscopic recombinations which have the potential to alter our mental maps of Europe (Wolff, 1994). The phenomenon is manifested by an upsurge in international institutional activity in Central and Eastern Europe. This section provides an overview of various integration initiatives, and lays the foundation for the subsequent examination of the effects of subregional coalescence.

Subregional and regional integration initiatives may be divided into formal and informal institutional arrangements. A formal initiative is an arrangement which imposes legal obligations on the contracting parties and creates institutions to carry out the purposes of the agreement. Informal integration initiatives are arrangements involving creation of institutions for more general cooperative purposes such as discussion of common interests and undertaking joint projects, but which do not actually impose legal obligations on the members.

The designations formal and informal do not correspond to a particular governance structure. In fact, nearly all developments considered here have an intergovernmental (rather than a supranational) character. They are, for the most part, facilitative meso-institutions (Abbott and Snidal, 1995) that have been 'created merely to facilitate the parties' integration goals' and are not 'actually empowered to produce substantive integration results in forms such as new norms, dispute resolution decisions, and harmonization legislation' (Garcia, 1997, p. 360).

Formal integration initiatives

Current formal integration projects include, at the regional level, the EU and the CIS, and at the subregional level, the Central European Free Trade Area, the Baltic Free Trade Area, and the Central Asian Economic Community, some of which are described below.

Central European Free Trade Area

The prime example of viable subregional economic integration in Central/Eastern Europe is the Central European Free Trade Area (CEFTA), which has been in effect since March 1993 (Bárta and Richter, 1996; Hartnell, 1997, pp. 181–9). CEFTA is the fruit of coordination among Czechoslovakia, Hungary and Poland, which started at their meeting in Visegrád in early 1991 (Vachudova, 1993). The Visegrád group of countries perceived their cooperation as 'a new pattern of relations in Central Europe', and the formation of CEFTA 'expressed [the] will to stop the disintegration of the links between their economies and [the] trust in the beneficial influence of strengthened cooperation on the economic growth and welfare' of the members' economies (Lawniczak, 1996, p. 139).

CEFTA is poised to undergo geographic expansion. Slovenia joined in January 1997, and was followed by Romania in July 1997, which brought the market up to 67 million people. Bulgaria joined in early 1999, and there is strong support for allowing the Baltic States to join (Potroro Summit Declaration, 1997). Other countries, including Croatia and Ukraine, have expressed interest in prospective CEFTA membership when they can fulfil the preconditions (which include WTO membership). CEFTA's expansion reflects the improved trade results and internal dynamics of CEFTA, but also reveals CEEC perceptions both of the advantages of participation in a subregional integration initiative, and of the diminishing likelihood of quick accession to the EU. For those countries with which the EU has already commenced accession negotiations, CEFTA is clearly a way-station, though one in which they are likely to stay for many years to come. As for other EA and perhaps also PCA countries, CEFTA will remain a feature of the New European landscape (Lechmanova, 1998).

The CEFTA Agreement provides for the gradual establishment of a free trade area for industrial goods. The weak intergovernmental institutional structure provides neither general legislative competence, nor any judicial authority empowered to make binding interpretations of the Agreement. And yet the economic and political importance of CEFTA in European integration should not be underestimated.

CEFTA illustrates a number of the characteristics and potential advantages of coalescence via subregional institutions. First, despite their early fear of being lumped together by the EU and treated as an amorphous mass, the Visegrád countries gradually overcame their reluctance to raise their common voice. Even the Czechs and the Slovenes, who initially tried to circumvent subregional cooperation, have come to embrace CEFTA, though still more pragmatically than passionately. It is this willingness to cooperate for mutual gain – rather than formal organizational strength – that accounts for CEFTA's growing importance in the region.

Second, CEFTA illustrates a number of dynamic structural features. For example, member states can (and do) propose to add new areas to CEFTA's competence. Also, CEFTA permits members to pursue closer relations with one another, as well as with non-member countries, provided that these do not negatively affect the CEFTA trade regime or rules of origin. CEFTA countries have in fact concluded many bilateral trade agreements with other CEEC countries. This criss-cross pattern of bilateral linkages has the potential to contribute towards further integration in CEFTA itself, since the CEFTA Agreement calls upon the members to 'to examine in the Joint Committee the possibility of extending to each other any concessions they grant or will grant to third countries with which they conclude a Free Trade Agreement or other similar agreement'.

Subregional success, measured in both economic and political terms, ensures that CEFTA is not only here to stay, but could well increase in importance in the coming years. Since all current members of CEFTA are EA countries, CEFTA can serve as their voice *vis-à-vis* the EU. CEFTA offers continuing advantages to the handful of countries which may be expected to drop out of CEFTA upon their accession to the EU, not least since enlargement will proceed slowly. Further, CEFTA offers a framework within which the fast-track CEFTA members can assist other CEFTA members in their dealings with the EU.

Baltic Free Trade Area and customs union

The Baltic States have devoted a great deal of effort to achieving closer integration among themselves, in addition to seeking closer ties with the EU, CEFTA and other CEECs. They established a Baltic Council of Ministers, concluded free trade agreements for both industrial and agricultural goods, and have been working towards establishing a customs union among themselves. However, there has also been considerable friction among these countries, largely in the context of their respective relations with the EU, which was exacerbated by the EU's 1997 decision to place Estonia (but not Latvia or Lithuania) on the accession fast-track.

Informal subregional integration

There has been a virtual explosion of informal subregional integration institutions having a more or less geographic identity. It is common for a given country to belong to more than one of these cooperative initiatives, as it is for each initiative to claim a diverse membership drawn from the EU, the CIS, Central and Eastern Europe, and (occasionally) elsewhere. Some informal institutions are concentrated in a geographic area. Overall, the informal institutional initiatives overlap in terms of membership, and transcend usual boundaries. These informal organizations have become audible voices within the cacophonous European house, despite the fact that they tend to be even weaker than the formal entities, both substantively and institutionally.

Council of Baltic Sea States

The Council of Baltic Sea States (CBSS, established in 1992) provides a good example of the loose type of structure that is proliferating in Europe, and a good contrast to the more focused CEFTA institutional model. Its membership includes the Baltic States (Estonia, Latvia and Lithuania), some EU countries (Denmark, Germany and Sweden), some non-EU countries (Iceland, Norway, Poland and Russia), and even the European Commission itself. The overriding goal of the CBSS is the development of 'good-neighbourly relations' among its members, which includes 'political contacts, and development of trade and cooperation in all spheres of life' (Gdansk Communiqué, 1995).

Black Sea Economic Cooperation

The Black Sea Economic Cooperation (BSEC) was established in 1992 by Albania, Armenia, Azerbaijan, Bulgaria, Georgia, Greece, Moldova, Romania, Russia, Turkey, and Ukraine, as a subregional structure for multilateral cooperation covering a broad range of economic and other activity. It is extraordinary that BSEC continues to exist, in light of the high degree of tension (and even conflict) among some of its members. Indeed, BSEC's tenacity in the face of these obstacles further affirms the attractiveness of subregional initiatives as a means to pave the way to integration into the global economy and to improve the bargaining position with the West (and in particular the EU).

Central European Initiative

The Central European Initiative (CEI, established in 1989) started out as a loose coalition (under Italian leadership) that from time to time has brought together representatives of Albania, Austria, Belarus, Bosnia-Herzegovina, Bulgaria, the Czech Republic, Italy, Hungary, Macedonia, Moldova, Poland, Romania, Slovakia, Slovenia, and Ukraine. CEI, with its permanent secretariat in Trieste, is akin to an interregional 'structured dialogue' with a fluid agenda and floating participation, consonant with its design as a 'flexible and pragmatic tool' for overcoming divisions in Europe (Cremasco, 1992). CEI is notable for its attempt to redefine its subregion as a centre, and thereby to subvert its marginalization. Despite considerable problems facing its members, the CEI appears to be flourishing and even expanding its scope of activities. Indeed, it has dared to address some of the more intransigent problems, such as Balkan security and agriculture. Rather like CEFTA, which is in some respects its formal counterpart, CEI serves now and then as the voice for its members in other international fora.

Interregional linkages

The current state of East–West European relations is characterized by a high degree of formal and informal interregional linkage. These proliferating initiatives are detailed elsewhere (Hartnell, 1997, pp. 209–12). The thickening interregional web contributes to the ongoing process of decentralization of power and process, gradual levelling of the trade regime, and disintegration of boundaries. Such linkages are occurring even in settings riven by contention, such as the Balkans and Transcaucasia, where coalescence is slowly overcoming even the fiercest old boundaries (see, e.g., Fuller, 1997).

Fringe benefits

Taken as a whole, these formal and informal initiatives are affecting the ongoing process of European integration in unexpected and salutary ways. The trend towards subregional (and to a lesser degree interregional) coalescence creates a counterweight to the power of the big regional initiatives. Thus the spokes are linked at the rim as well as at the hub, and the rim grows stronger. In this sense coalescence may be said to lend stability to Europe. But this stability is anything but rigid: coalescence implies fluidity, adaptability, shifting and overlapping boundaries, multiple hubs and intersecting gears. It is a three-dimensional, rather than a flat-earth, view of Europe. Indeed, coalescence means migration of the centre and redrawing of lines. These momentous shifts do not, however, obviate the existing vertical integration framework established by the EA agreements. Rather, the recent changes in Central/East European consciousness manifest themselves partly in incremental corrections of some of the pathologies identified above. Subregional coalescence not only begins to redress some of the imbalances present in the EU's top-heavy vertical integration model, it also influences the European integration agenda.

Economic benefits

The success of subregional coalescence can be measured in the economic sphere. Despite CEFTA's sluggish beginnings, trade has increased significantly among member states since 1994, which partly explains why other CEECs have clamoured to join. This is a success that can be enjoyed by the EU, which pushed the Visegrád group of countries to rebuild their trade ties, as well as by the CEFTA member states themselves, which are reaping the immediate benefits. On the road to development, therefore, subregional integration is an effective motor, albeit no substitute for the CEECs' overall integration into the global trading system. Indeed, CEFTA's goals are to 'gear up' the subregional economies and thereby improve their chances of merging onto the EU autobahn, whilst compensating for lack of access to European markets in the meantime.

The economic benefits of subregional cooperation – particularly within the CEFTA framework – do not stop at increased trade among member states. Coalescence has ameliorated the trade pathology by narrowing the gap between the EU and the EA countries. By joining their voices the CEFTA countries have been able to influence the EU's policy towards them within the EA association framework. For example, the CEFTA countries pushed consistently and successfully to obtain (gradually) improved access to the EU market, especially for sensitive exports such as textiles, agricultural and ECSC products. Further, the EU has implemented some of the

CEFTA countries' proposals to reform the PHARE program, and finally implemented a system of cumulative rules of origin.

Another benefit of subregional coalescence is that CEEC (and particularly CEFTA) countries have recognized the possibility of integrating further and faster at subregional level than at regional level (e.g., within the EU's association framework). CEFTA member states have actively considered whether to expand their agreement to cover a variety of matters – such as trade in services, free movement of capital, transport, telecommunications, energy, infrastructure, and privatization – that are beyond the scope of their EA treaties with the EU. The CEFTA Agreement could thus surpass the scope of the EA agreements that served as its template. CEFTA member states could also move beyond mere trade liberalization, and cooperate in ways that outstrip the integration achieved by the CEECs under their respective EAs with the EU, as the Baltic States have already done, e.g., in regard to agricultural trade. Such subregional integration could drive further integration at the regional level, and thereby alter the course of the top-down integrative forces emanating from Brussels.

Subregional economic integration does more than just further the cause of transition in Central and Eastern Europe. It demonstrates that European integration can proceed not just vertically, from the top down and according to an agenda set by the EU (or by Moscow), but upon multiple fronts, and at different levels, corresponding to different needs. Jacques Delors aptly equated regionalization (and, by extension, subregionalization) with 'worldwide subsidiarity' (Delors, 1995, p. 724). The existence of integration initiatives at the lower end of the food chain evokes the prospect of integrative pressure from the bottom up.

Political benefits

The coalescence that results from the proliferation, interaction, and maturation of formal and informal subregional integration initiatives has significant political implications for European integration. It has nurtured self-confidence and sophistication in the countries of Central and Eastern Europe, in addition to demonstrating the benefits of subregional cooperation. These effects are visible in the relations of the CEECs to one another and in their relations to the big regional powers in Europe. But the effects of coalescence can also be seen outside Europe, as in the 1993 discussions regarding CEFTA–NAFTA economic cooperation.

It bears repeating that the formal and informal initiatives in Central and Eastern Europe are all weak, intergovernmental institutions, without power to bind member states in the absence of their consent. Still, there is evidence of growing confidence in the cooperative process, if not also in the new institutions themselves.

Coalescence directly affects the relations between the EA (and particularly the CEFTA) countries and the EU. The current level of cooperation must be measured against CEFTA's early years, when the member states scrupulously avoided developing a common voice or otherwise expanding the scope of their cooperation. At that time, each CEFTA country was preoccupied with asserting itself individually *vis-à-vis* the EU, and with trying to distance itself from the pack in the race to accession. Today CEFTA countries still insist that the EU consider each of them on their individual merits in accession negotiations ('differentiation'), but they no longer hesitate to assert themselves as a group.

In the context of relations between the EU and the associated countries in Central and Eastern Europe, coalescence has ameliorated the trade, aspirational, participation, and sovereignty pathologies, though the relations still suffer from considerable '*dis*-ease'. Payoffs from subregional cooperation are visible in the negotiated improvements to the EA framework. EA countries would be wise to weigh the benefits of further cooperation (or at least coordination) as they enter into accession negotiations, particularly in the light of their weak position *vis-à-vis* the EU, which weakens further as the transition period for full implementation of each EA agreement's provisions on the movement of goods nears its end. Once the free-trade areas called for by the respective EA agreements are in place, the EU has achieved its pan-European free trade area, and might not see fit to make further substantial financial commitments to the CEECs. Yet from the perspective of the EA countries, their agreements to expose their economies 'to the ravages of open competition' were secured by the prospect (if not the promise) of 'some sort of payback under the [EU's] structural, regional or cohesion funds programmes' ('Long Sprint', 1997, p. 40). The EA countries are in a precarious position, since they must fully implement the EAs to stand a chance of succeeding in accession negotiations with the EU.

At the outset of negotiations with applicant countries, the EU is expected to offer a fairly standard set of accession conditions to all, and then rely on parallel negotiations with each EA country to tailor a suitable compromise. In this scenario, Bárta and Richter (1996, p. 29) predicted that 'diverging individual CEEC positions on important issues may negatively influence CEEC bargaining power: the less demanding CEEC negotiators could provide precedents for the lowest [set of accession criteria] to which EU negotiators could refer in other bilateral relations. Thus the "softest" of the CEECs may lower the chances [for] the less conciliatory ones.'

Richter (1997, p. 22) has called for 'joint behaviour rules' and even for a 'joint accession strategy'. Joint behaviour rules could serve, at the very least, to prevent non-coordinated concessions by one or more applicant countries. Signals from fast-track EA countries concerning readiness to coordinate their accession negotiations have been mixed, but growing stronger. On a practical level the chief accession negotiators of the five 'fast-track' CEEC countries have been coordinating their work, intend to continue doing so, and have decided to expand their cooperation by creating multinational groups of experts to work on various aspects of the *acquis* (*Agence Europe*, 14 October 1998, pp. 8–9), despite being warned by the EU that a coordinated approach might prove harmful. This multilateral tendency may begin to compensate for the lack of any meaningful political opposition to EU accession at home (Bárta and Richter, 1996, p. 20).

The cooperative spirit does appear, however, to be beset with contradictions, if not limitations. The talk of politicizing CEFTA has been especially controversial. Hesitations and setbacks notwithstanding, subregional cooperation will play an ongoing role in the context of accession negotiations with the EU, whether through CEFTA, the revitalization of Visegrád (*or both*).

The relations between the EU and the aspirants in Central and Eastern Europe (in particular the CEFTA member states) have been characterized by frustration on both sides. The Visegrád countries (which constitute the core of CEFTA) pushed the EU hard for concrete commitments on eventual membership. The EU, caught between its good intentions, on the one hand, and its own internal political and economic difficulties, on the other, has been forthcoming, although not enough to alleviate the pressures on the governments in the associated CEECs, which therefore continue their vigilant efforts to persuade the EU to take them more seriously. Since 1997 fast-track EA countries have voiced deep longing for partnership with the EU in lieu of patronization. Subregional coalescence responds to this need.

One of the most welcome consequences of subregional integration is that Central and Eastern European countries which lack adequate participatory rights within the EU's association framework, have created forums in which they do have full participatory rights, and where they can play a role in setting the European integration agenda. This is not to overstate the case by asserting that the tail is firmly wagging the dog, since all formal and informal initiatives involving EA countries are intended to further the aim of joining the EU as soon as possible. Still, these new subregional institutions are developing independent values of their own, which reflects a real turn of the tide.

The EU is no longer fully in charge of the agenda of European integration, and agenda-setting has become multipolar. The EU realized as early as 1995 that it should play a more active role in some of the new informal subregional initiatives (e.g. the Council of Baltic Sea States and the Central European Initiative), and commenced sending top-level representatives to key meetings. Still, the EU is the regional hegemon, the 'regime maker' in Europe, whose actions (and inactions) profoundly affect the transformations in Central and Eastern Europe. The CEECs have learned from experience how to negotiate with the EU; their elites are socialized and their preferences shaped through exposure and apprenticeship to the EU institutions (Nikolaïdis, 1993). In this world, subregional coalescence gives marginal actors leverage with which to move the big regional trade blocs, however incrementally. While no major power shift is imminent, even a small impetus can be turned to real advantage by strategic actors firmly rooted in their marginal positions.

The competition–cooperation paradox is likely to be resolved by increasing numbers of CEECs in favour of strategic cooperation, owing both to growing CEEC disappointment in the EU, and successful subregional cooperation in economic and political spheres. And yet the competition–cooperation paradox is sure to retain significant (and possibly even greater) force for some CEECs, largely because of the EU's decision to create a two-tier enlargement process. Thus, while the 'in' CEEC countries see fit to cooperate, at least with some members of their group, competition between 'in' and 'out' countries, as well as among 'out' countries, will increase. In the face of this pernicious divisive situation, and of the certain delays ahead on the road towards union, the 'ins' would do well to remember, and build upon, what they have already achieved through strategic cooperation.

Conclusion

This chapter has examined the coalescence of subregional integration initiatives in the European context, with a particular focus on how this phenomenon affects the politically charged process of European integration. Some economists consider regionalism and subregionalism undesirable because they result in trade diversion and inefficiency (Bhagwati and Krueger, 1995; see also Mansfield and Milner, 1997). Whether or not those economists are right, regionalism and subregionalism offer further economic and political benefits that are sorely needed in Central and Eastern Europe. Subregional integration is no substitute for further integration at the regional (or for that matter global) level. But coalescence can ameliorate the pathologies of the EU's regional top-down, vertical integration model, and serve the cause of further, fairer European integration.

I subscribe to Jacques Delors' view that regionalism (and thus subregionalism) 'paves the way to a more coherent and more legitimate international order' without 'abandoning cultural roots' (Delors, 1995, pp. 716–23). Further, I believe that subregional coalescence is necessary if European integration is to rise to Delors' challenge that it respect common principles and strive towards a more 'legitimate, coherent form'. Salutary developments on Europe's margins are pushing the EU in this direction, though it is far from ready, especially in hard economic times, to face the changes that will be necessary in the future. The EU has promised membership, and some day it will have to deliver.

Meanwhile, European integration consists of multiple processes that must be viewed in relation to each other. They frequently influence, occasionally complement or conflict with, often overlap with, and sometimes even nest within one another (Aggarwal, 1998). A multipolar vision of Europe that includes marginal perspectives requires us to abandon the telescopic, distancing visions of concentric-circle or hub-and-spoke integration, in favour of a kaleidoscopic one, which mixes and recombines the many elements of Europe's beauty.

References

Abbott, Kenneth and Snidal, Duncan (1995) 'Mesoinstitutions: the Role of Formal Organizations in International Politics' (unpublished manuscript).
Aggarwal, Vinod (1998) 'Reconciling Multiple Institutions: Bargaining, Link-

ages, and Nesting', in Vinod Aggarwal (ed.), *Institutional Designs for a Complex World: Bargaining, Linkages, and Nesting* (Ithaca, NY: Cornell University Press), pp. 1–31.

Ágh, Attila (1998) *The Politics of Central Europe* (London: Sage).

Bárta, Vít and Richter, Sándor (1996) 'Eastern Enlargement of the European Union from the Western and from the Eastern Perspective' (unpublished manuscript).

Bhagwati, Jagdish and Krueger, Anne (eds) (1995) *The Dangerous Drift to Preferential Trade Arrangements* (Washington, DC: AEI Press).

Bielski, Stefan (1995) 'To Be Or Not To Be Peers', *Warsaw Business Journal*, **1**(30).

Budge, Ian, Newton, Kenneth and McKinley, R. D. (1997) *The Politics of the New Europe: from the Atlantic to the Urals* (London and New York: Longman).

Cremasco, Maurizio (1992) 'From the Quadrangolare to the Central European Initiative – an Attempt at Regional Cooperation' in A. Clesse *et al.* (eds), *The International System after the Collapse of the East–West Order* (Dordrecht: Martinus Nijhoff), pp. 448–59.

Daedalus: Journal of the American Academy of Arts and Sciences (1997) 'Preface to the Issue "A New Europe for the Old?"', **126**(3), v–xii.

Delors, Jacques (1995) 'The Future of Free Trade in Europe and the World', *Fordham International Law Journal*, **18**, 715–25.

Dupont, Cédric (1998) 'The Failure of the Next-Best Solution: EC–EFTA Institutional Relationships and the European Economic Area', in Vinod Aggarwal (ed.), *Institutional Designs for a Complex World: Bargaining, Linkages, and Nesting* (Ithaca, NY: Cornell University Press), pp. 124–38.

European Dialogue: The Magazine for European Integration <http://europa.eu.int/en/comm/dg10/infcom/eur_dial/index.html>.

Fuller, Liz (1997) 'New Geo-Political Alliances on Russia's Southern Rim', *RFE/RL Newsline* <http://www.rferl.org> (16 April).

Garcia, Frank (1997) 'New Frontiers in International Trade: Decisionmaking and Dispute Resolution in the Free Trade Area of the Americas: an Essay in Trade Governance', *Michigan Journal of International Law*, **18**, 357–97.

'Gdansk Communiqué', *Fourth Ministerial Session of the Council of the Baltic Sea States*, <http://www.baltinfo.org/> (18–19 May 1995).

Hadas, Miklós and Vörös, Miklós (eds) (1997) 'Ambiguous Identities in the New Europe', *Replika: Hungarian Social Science Quarterly*, Special Issue.

Haggard, Stephan, Levy, Marc A., Moravcsik, Andrew and Nikolaïdis, Kalypso (1993) 'Integrating the Two Halves of Europe: Theories of Interests, Bargaining, and Institutions', in Robert O. Keohane *et al.* (eds), *After the Cold War: International Institutions and State Strategies in Europe, 1989–1991* (Cambridge, MA: Harvard University Press), pp. 173–95.

Hartnell, Helen (1993a) 'Central/Eastern Europe: The Long and Winding Road toward European Union', *Comparative Law Yearbook of International Business*, **15**, 179–229.

Hartnell, Helen (1993b) 'Association Agreements between the EC and Central and Eastern European States', *Acta Juridica Hungarica*, **35**, 225–36.

Hartnell, Helen (1997) 'Subregional Coalescence in European Regional Integration', *Wisconsin International Law Journal*, **16**(1), 115–226.

Judt, Tony (1997) *A Grand Illusion? An Essay on Europe* (London: Penguin).

Kennedy, David (1991) 'Turning to Market Democracy: a Tale of Two Architectures', *Harvard International Law Journal*, **32**, 373–96.

Kolankiewicz, George (1994) 'Consensus and Competition in the Eastern Enlargement of the European Union', *International Affairs*, **70**, 477–96.

Lawniczak, Ryszard (1996) 'Sub-Regional versus Pan European Free Trade of the CEECs', in F. Laursen and S. Riishøj (eds), *The EU and Central Europe: Status and Prospects* (Esbjerg: South Jutland University Press), pp. 139–60.

Lechmanova, Natalie (1998) *Central European Dilemma: EU or CEFTA Integration* <<http://www.faterni.com/conferences/lechmanova.html>>.

Lengyel, László (1995) 'Towards a New Model' (electronically published manuscript).

'Long Sprint' (1997) *Business Central Europe*, February, pp. 38–40.

Malia, Martin (1997) 'A New Europe for the Old?', *Daedalus: Journal of the American Academy of Arts and Sciences*, **126**(3), 1–22.

Mansfield, Edward and Milner, Helen (eds) (1997) *The Political Economy of Regionalism* (New York: Columbia University Press).

Mayhew, Alan (1998) *Recreating Europe: The European Union's Policy towards Central and Eastern Europe* (Cambridge: Cambridge University Press).

Moravcsik, Andrew (1993) 'Preferences and Power in the European Community: a Liberal Intergovernmentalist Approach', *Journal of Common Market Studies*, **31**(4), 473–524.

Nikolaïdis, Kalypso (1993) 'East European Trade in the Aftermath of 1989: Did International Institutions Matter?', in Stanley Keohane *et al.* (eds), *After the Cold War: International Institutions and State Strategies in Europe, 1989–1991* (Cambridge, MA: Harvard University Press), pp. 196–245.

Peers, Steve (1995) 'An Ever-Closer Waiting Room? The Case for Eastern European Accession to the European Economic Area', *Common Market Law Review*, **35**, 187.

Pocock, J. G. A. (1997) 'What Do We Mean by Europe?', *The Wilson Quarterly*, Winter, pp. 12–29.

'Potroro Summit Declaration' (1997) *Conclusions of the Summit of the Prime Ministers of the CEFTA Countries* <http://www.ijs.si/cefta/eng/index.html-12> (12–13 September).

Richter, Sándor (1997) 'European Integration: the CEFTA and the Europe Agreements', *WIIW (Wiener Institut für Internationale Wirtschaftsvergleiche) Research Reports*, no. 237 (Vienna, May).

Rushdie, Salman (1996) 'In Defense of the Novel, Yet Again', *The New Yorker*, 24 June, p. 48.

Shinkarenko, Pavel (1996) 'Five Years after the Fall: CIS is Alive, if not Kicking', *Eastern Europe New Digest* <europe-news.request@hookup.net> (10 December 1996).

Vachudova, Milada A. (1993) 'The Visegrad Four: No Alternative to Cooperation?', *RFE/RL Research Report*, **2**(34), 38–47.

Wilson, Kevin and van der Dussen, Jan (eds) (1995) *The History of the Idea of Europe* (London: Routledge).

Wolff, Larry (1994) *Inventing Eastern Europe: The Map of Civilization on the Mind of the Enlightenment* (Palo Alto, CA: Stanford University Press).

3

European Integration, Market Liberalization and Regional Economic Disparities

Valerio Lintner

Introduction

The distribution of economic wealth and of economic activity across Europe, and how this is affected by economic (and political) integration within the European Union, is a fundamental parameter which impinges on virtually all aspects of this book. For comparisons of wealth or indications of economic dependency are amongst the most obvious signs of marginal status. Some examination of this basic issue is therefore required. The essential aim of this chapter is to examine, within an interdisciplinary (but predominantly economic) framework, the likely effect of European integration on the regional distribution of economic welfare. Put another way, it sets out to discuss whether European integration increases regional disparities or narrows them. If the former, then the EU's integrated market will tend to spawn margins and encourage the tensions of marginality; if the latter, integration will reduce marginality and its effects. The chapter can be seen as essential background to other parts of the book, as well as dealing with a subject of great importance in its own right. Apart from being central to any discussion of regional development and regional policy, it also offers important insights into the likely effects of integrating economies and societies at different phases of economic development, which is just what is envisaged in the further enlargement of the EU.[1] If forthcoming enlargement is to promote *greater* regional disparities, it will bring with it heightened tensions on the margins of integration.

First, the chapter reviews the debate on integration and regional disparities, concentrating on the very fundamental theoretical and policy question which underlies the debate: What is the impact of the liberalization of markets on regional activity and distribution? Then it reviews the empirical work in the area, and offers some conclusions, including a few considerations on the development, and the appropriateness of EU regional policy.

Theoretical debate: neo-classical positions

The impact of integration on regional distribution is a relatively under-researched area, at least as far as economics, and in particular mainstream economics, is concerned. Other disciplines, such as economic geography, have considered the issue more fully, while important insights into the area have also been offered by the social scientists of the Regulation school.[2] According to the latter, inter-disciplinary, approach to the analysis of how societies function, economic success broadly occurs as a result of the promotion of a harmonious relationship between basic factors such as technology, the nature of work, economic policies and relations between the major social actors: labour, capital and the state. The latter plays a key role, being largely responsible for providing the mode of regulation, or the 'rules of the game' within which social and economic conflict is mediated, as well as some of the means of mediating social tensions, for example welfare provision. It is only when conflict can be successfully mediated that high rates of economic growth and accumulation can occur.

This approach would suggest that successful areas will broadly be those that manage to create the institutional, policy, regulatory and other mix which is required to minimize conflict between the various social actors, thus creating the conditions in which the local economy can prosper.

The neo-classical economic theory of integration[3] largely ignores issues of distribution, which are not really considered to be of importance from a technical point of view. The process of integration results in there being gainers and losers, but as long as it can be demonstrated that 'the gainers gain more than the losers lose' all is well. For it is then possible for the gainers to compensate the losers and still be better off themselves. The analysis tends, that is to say, to concentrate on the *overall* effect of policies and developments. It is left to the political process to engineer any redistribution. The problem is that in the real world gainers rarely compensate losers, and so regional (and other) distribution does in fact constitute an important issue.

On the face of it, European integration can affect the regional distribution of welfare and economic activity in a number of ways:

- the customs union involves changes in protection levels, which result in trade changes[4] that in turn reallocate productive resources and thus affect the distribution of regional income;
- the intra-EU free mobility of labour and capital also involves a reallocation of resources and economic activity which is unlikely to be neutral at the regional level;
- the common policies of the EU, and in particular the CAP, have significant regional effects;
- the structural funds redistribute regional income directly;
- monetary integration will have further liberalizing and therefore reallocating effects: for example the 'transparency of prices' produced by the single currency, reduced transaction costs, and reduced uncertainty for those involved in trade will all result in changes in the location of economic activity. This will have a regional dimension,[5] just as has been the case with earlier measures of integration.

From a more theoretical perspective, one could argue that in essence integration affects regional income distribution in three ways:

1. By depriving national governments of potential policy tools for influencing regional distribution: e.g. tariff and non-tariff barriers (such as subsidies, capital controls, and interest rates); fiscal policies and exchange rates (in the case of the Maastricht model of EMU and the Stability and Growth Pact which goes with it).[6] However, it should be borne in mind that the transfer to the supranational level of policy-making involved in integration can in some cases *increase* the *effectiveness* of policy. This resembles the argument for there being an increase in sovereignty as a result of 'pooling' it. It may be the case that EU policies work better than the national or regional ones that they replace, and macroeconomic policy within EMU may be an instance of this. The regional dimension of this is nevertheless a matter for debate.
2. By the direct effects of the EU's redistributive policies. Though this may be important in particular sectors in a microeconomic sense, given the limited size of the EU budget it is unlikely to be of great macroeconomic or redistributive significance. The MacDougall report (European Commission, 1977), which was commissioned as part of the first attempt at monetary union

(the Werner Plan) suggested that, at that time, a budget of 7 per cent of Community GDP would be needed to eliminate 40 per cent of existing inequalities. Times and the EU may have changed; but the existing *total* of the EU budget remains a mere 1.27 per cent of EU GDP. Even the modest proposal of an increase to 1.38 per cent was rejected at the Edinburgh summit.[7]

3 By freeing the operation of markets. Given the broadly neo-liberal development of the EU and the global economic climate prevailing over the past two decades, the process of economic integration in Europe has invariably involved the progressive liberalization of markets and of exchange both within EU countries and between them – that is to say, deregulated *laissez-faire* economies. The customs union and common market aimed to create an internal 'level playing field' in trade in goods and services as well as in the allocation of labour and capital resources. EMU will reinforce the liberal character of the common market by reducing uncertainty and transaction costs and, as mentioned above, by promoting the 'transparency of prices'. It is important to note, however, that the liberalization of exchange and increased capital mobility are well-established elements of globalization. Though the EC/EU has certainly contributed to their development, they probably would have happened even in the absence of the EC/EU integration.

Given what has been said already, in the long run market liberalization is likely to be the most significant factor linking economic integration with relative regional development. This is not a proposition which can be tested empirically, at least not in a strict quantitative sense, but it is borne out by the author's view of the world. There are, however, a couple of caveats. First, as mentioned above, it is difficult to determine the extent to which the EU, as against an established global trend, is actually responsible for liberalization of markets. It may be that this would have occurred even in the absence of the EC/EU. What is more, it should be remembered that customs unions (and indeed common markets) involve elements of protection as well as of liberalization. Trade diversion due to barriers around the zone may result in losses of consumer welfare in some regions through higher prices; but it also promotes local economic activity because increased protection favours domestic industries *vis-à-vis* third-country producers.[8]

The key question is therefore whether the free or freer operation of markets which results from economic integration tends to narrow regional disparities – i.e. producing convergence – or to increase them – i.e., producing divergence. We shall therefore concentrate on that question. Needless to say, there are differing views on this key issue, which goes directly to the heart of the debate on the nature of markets themselves, a debate itself fuelled by the difficulties of empirical analysis referred to above.

Traditional models of the effects of market integration – such as those of Borts (1960) and Borts and Stein (1964) – have reflected neo-classical views on markets. The same could be said of more recent 'catch-up' models, such as Barro and Sala-i-Martin (1991, 1992). Fundamentally, these are based on the theoretical proposition that the free operation of markets tends to promote factor-price equalization; that is, that the prices of the factors of production in different locations tend to move closer to each other.[9] In particular, there will be a convergence of wages and of rates of return on capital, and this leads to a convergence of economic conditions in the different geographical areas. To the extent that European integration opens up markets, it can thus be expected to favour regional convergence across the EU.

In neo-classical terms, convergence should occur essentially as a result of movements of labour and capital. Workers in poorer regions will tend to move to richer areas in response to higher wages and better employment opportunities. This reduces the supply of labour and pushes up wages in less developed regions, while restraining earnings in better-off areas. As far as capital is concerned, the creation of the common market results first in a once-and-for-all movement of capital from areas of low productivity to *more* productive, and hence also *more prosperous*, areas where the marginal product of investing it is greater. In due course, by increasing the supply of capital in the prosperous areas, this shift will reduce rates of return in these areas. It will also increase rates of return in poorer areas because of the reduction in the supply of capital, thus tending to equalize returns to capital across areas of high and low productivity. Thereafter, there will be a tendency for capital to move to *less productive, less prosperous* areas in response to lower labour and other costs there. This will raise wages and thus living standards in the poorer regions. This approach can be conveniently illustrated by reference to Figures 3.1 and 3.2.

Figure 3.1 depicts in a simplified form the effects of market liberalization on labour markets in two areas with different levels of economic prosperity. In this illustration the left-hand segment (region S) is

Figure 3.1 Neo-classical approach: convergence in regional labour markets

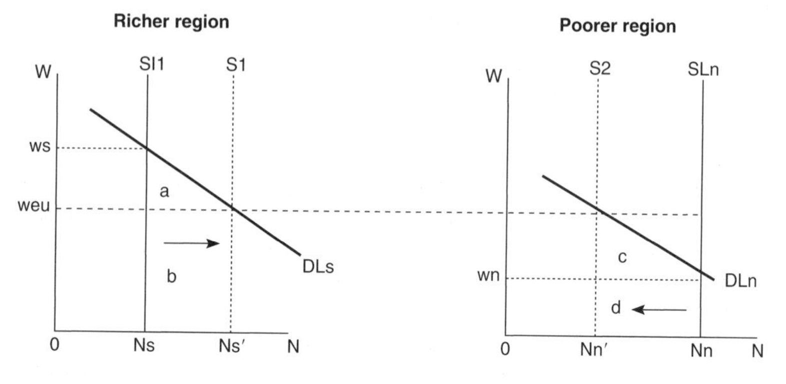

Figure 3.2 Factor price equalization or polarization? The effect of the free movement of capital

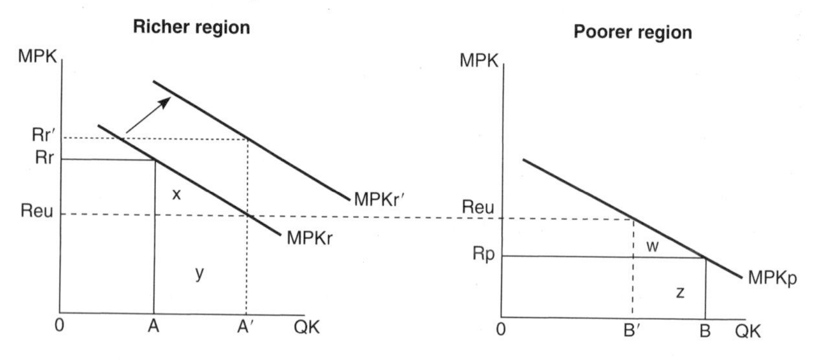

shown as the more well-off region, while the right-hand segment (region N) represents the poorer one. W on the vertical axes represents wage rates in both areas, N on the horizontal axes represents levels of employment, DLs and DLn are the demand curves for labour in each area (reflecting respective labour productivities), while SLs and SLn represent the supply of labour curves, shown for ease of illustration as unresponsive to changes in the wage rate. In the absence of free markets and labour mobility, the equilibrium wage in region S will be O-ws and in region N it will be O-wn. Employment in the respective regions is O-Ns and O-Nn. Free markets and mobility result in migration from N to S in response to the wage differential.[10] As labour leaves N the supply of labour curve shifts to the left and the wage rate in the region increases. Similarly wages fall in S in response to the influx of new workers, which shifts the supply of labour curve to the right. The process ends when wages are equalized at O-weu with Nn-Nn′ (which is equal to the influx Ns-Ns′ in region S) workers migrating, shifting the supply of labour curve in region S to S1, and in region N to S2. Thus the model suggests that the free operation of markets will result in the equalization of earnings and thus living standards between two regions.

Figure 3.2 looks at capital flows, which are likely to be more important in this context as labour is typically much less mobile than capital. This shows a simplified model of the effect of capital movements between two regions, again with different levels of economic wealth and development and different levels of productivity. The marginal productivity of capital (MPK) in each region is measured along the vertical axis, while the quantity of capital resources (QK) is measured along the horizontal axis. The left segment represents a relatively rich region with a higher level of capital productivity (MPKr) and a greater rate of return on capital in the absence of capital mobility (O-Rr with an endowment of indigenous capital O-A) than in the poorer region, in which the marginal productivity of capital is represented by the curve MPKp and the rate of return on capital is O-Rp with an endowment of capital O-B. The neo-classical position would be that, with the liberalization of markets, capital will move from the poorer region to the richer region in response to the differential in the rate of return. As capital leaves the poorer region the rate of return there increases in response to the increased scarcity of capital, while in the richer region rates of return fall in response to the capital influx. The process continues until B-B′ capital (equal to A-A′) has moved and rates of return are equalized at O-Reu. Thus the neo-classical position on capital movements is again that the liberalization of markets results in convergence.

The neo-classical model also predicts an increase in production and *welfare* as a result of free factor movements. In Figure 3.2 capital mobility increases domestic product in the richer region by x + y, while precipitating a decrease of a lesser amount in the poorer region (w + z).[11] Once the equalization of rates of return on capital has taken place, there is likely to be a flow of capital back towards poorer areas, in response to factors such as high property prices and labour shortages in well-off areas, or wage differentials and the availability of plentiful unemployed or under-employed labour in poorer areas (since, as we have already pointed out, labour is in practice far less mobile than capital). Thus in this neo-classical world convergence is further facilitated.

In addition to the above, it is claimed that free markets will result in *technological* diffusion, with advanced techniques spreading from more developed regions to relatively less advanced ones. These benign effects would be compounded by the fact that poorer regions, which start from a lower base in terms of productivity and living standards, may be able to sustain higher rates of growth: they are less likely to experience the diminishing returns that result from the already high levels of capital accumulation in richer areas. Furthermore, free trade will result in poorer regions with plentiful labour being able to specialize in the production of labour-intensive products in which they have a comparative advantage. This will increase employment, exports and welfare in the region. Thus the basic message is that the market automatically tends to eliminate regional differences and to promote convergence.

This view is fundamental to the neo-liberal economic agenda: it provides the intellectual justification for limiting redistributive policies, including regional policy. Such policies are considered ineffective, wasteful and unnecessary for promoting regional development and convergence. The only forms of regional intervention that are countenanced by the extreme neo-liberal approach are those designed to promote growth by enabling the unhindered operation of markets in less well-off areas, for example the creation of Enterprise Zones in the UK. Less extreme versions of the neo-liberal approach may, however, accept that labour is in practice immobile, and that there may be impediments to the free flow of real capital investment between regions. Some limited form of regional policy may then be countenanced in the interests of promoting greater mobility.

Theoretical debate: alternative approaches

On the other hand, there is the view that market liberalization in Europe is likely to increase regional differences and divergence. It is based on the observation that contemporary Europe does not in practice conform to the neo-classical model, and that there are in fact forces in operation which prevent markets from functioning in the way described above. Over time capital may remain in the core of the EU despite wage differences and other factors. Selective emigration may render periphery areas less attractive, as may other factors such as lack of infrastructure or a lack of skills among the workforce and location.

In practice, markets may actually work in a perverse fashion, because there are likely to be continuing factors which reinforce the attractiveness of more developed areas. Prominent among these factors are economies of scale; the 'external' economies and thus reduced costs which occur where industries are concentrated in particular areas (so-called 'economies of agglomeration'), and the possibilities of increasing specialization through intra-industry trade. This has led to the view that the opening of markets may result in a 'polarization effect', an idea derived from Gunnar Myrdal's concept of 'cumulative causation'.[12] Opening markets leads initially to capital flows into more developed areas where rates of return are greater. However, this process leads to an *increase* in rates of return in developed areas, because of external economies such as those described above, which attracts even more capital from less developed regions, in turn increasing rates of return in developed areas, and so on. Put another way, productivity growth of a region depends on how fast its output increases, which depends on competitiveness in export markets, which in turn depends on productivity growth.

There is then a circular, dynamic process making the developed regions' growth self-sustaining and progressively *increasing* disparities between rich and poor areas. Divergence is further reinforced by the effects of selective migration of skilled labour towards richer regions. It is only partially mitigated by 'spillover' or 'trickledown' effects in the form of capital flows into poorer areas as a result of factors such as congestion. Similar principles are found in the models of Kaldor (1970) and Dixon and Thirlwall (1975) – much influenced by Verdoorn's Law, which emphasizes the role of scale economies and specialization in increasing productivity in already developed regions. More recent models have emphasized technological change or education and training as forces for cumulative causation and increasing inter-regional polarization.[13]

This alternative approach to how free markets influence the spatial distribution of economic activity can be illustrated by reference to Figure 3.2. Here, instead of rates of return to capital converging towards Reu (as is predicted in the neo-classical model), the influx of capital into the richer region from the poorer one increases productivity in the richer area. That shifts the curve for the marginal product of capital (MPKr) out to the right (MPKr'). This in turn increases the rate of return to capital (Rr'), which leads to further capital inflows from the poorer area, and so on. Finally, in roughly the same tradition are models, pioneered by Perroux,[14] which are based on 'growth pole theory', and which emphasize the role of particular growth industries and activities that act as an engine to growth by attracting other industries required to supply inputs.

Empirical studies and issues

There is considerable debate amongst academics and policy-makers over these issues of convergence and divergence and what role the process of European integration plays in them. As mentioned above, empirical studies have not resolved matters much, since the issues examined in this chapter are very difficult to test empirically, at least in a strict sense using standard parametric analysis. Casual empiricism would suggest that the polarization effect may well operate within the EU. There is a distinct core–periphery pattern to the spatial distribution of economic welfare within the EU, as is well documented by the EU's basic data on income, employment and migration,[15] and by the data presented in Table 3.1. In fact, regional disparities within the EU are far greater than in other comparable countries and organizations. They are twice as great as in the USA, for example. But then the USA has been an integrated entity with liberalized markets for a long time. Certainly, up to the rise of neo-liberalism in the 1980s, the assumption had been that European integration did in fact exacerbate regional disparities. It is only in the past 10–15 years that this assumption has been called into question. It is very difficult to test the convergence or divergence hypotheses partly because of the difficulties inherent in measuring welfare. There are a number of different economic definitions of welfare, and it can reasonably be argued that purely economic variables are not in any case an adequate measure of social welfare. In addition, there are the difficulties involved in isolating the effects of integration and market liberalization from the myriad of other factors which affect regional development. Nevertheless there is a body of empirical research in the area, and this section will first of all examine studies of overall convergence and divergence in the EU, and then those which attempt to isolate the effects of the process of integration.

If we look first of all at the issue of overall convergence and divergence, then the pattern of official indicators such as per-capita GDP and unemployment suggests that up to 1974 differences between

Table 3.1 Europe's regional prosperity league

Region	GDP per head (% of EU average)	Population (million)	% GDP Agriculture	% GDP Industry	% GDP Services
Wealthiest regions					
1. Hamburg (Ger)	196	1.7	0.3	23.2	76.5
2. Wallone (Bel)	183	3.3	0	19.2	80.8
3. Luxembourg	169	0.4	1.5	31.0	67.5
4. Ile de France	161	10.9	0.3	25.0	74.8
5. Bremen (Ger)	156	0.7	0.3	29.8	69.9
6. Hessen (Ger)	152	5.9	0.5	26.7	72.8
7. Lombardia (It)	131	8.9	1.9	37.6	60.6
8. { Bayern (Ger)	128	11.9	1.1	34.7	64.2
Emilia-Romagna (It)	128	3.9	4.6	33.3	62.1
10. { Ahven/Aland (Fin)	126	0.03	6.8	16.1	77.2
Baden-Wurten. (Ger)	126	10.3	1.0	40.5	58.6
12. Ostosterreich (Au)	122	3.4			
13. { Lazio (It)	119	5.2	1.8	19.0	79.2
Nord Est (It)	119	6.5	3.7	33.0	63.3
16. Nord Ovest (It)	116	6.1	2.6	31.8	65.6
17. Bruxelles (B)	115	0.9	1.9	32.6	65.5
18. Denmark	114	5.2	3.7	27.0	69.3
19. West-Nederland	113	7.2	2.9	21.1	76.1
20. Nord.-Westf. (Ger)	112	17.7	0.7	36.3	63.0
21. Westoster. (Au)	110	2.9			
22. Centro (It)	107	5.8	3.1	31.2	65.6
23. { Saarland (Ger)	106	1.1	0.3	34.4	65.2
Schleswig-Hol. (Ger)	106	2.7	2.2	29.6	68.2
Poorest regions					
60. Sicilia (It)	70	5.1	6.7	20.2	73.1
61. Campania (It)	69	1.6	3.8	20.3	76.0
62. { Continente (Por)	68	9.4	4.1	34.1	61.8
Sud (It)	68	6.7	7.2	19.8	72.9
64. Nisia A., Kriti (Gr)	67	1.0	23.5	15.4	61.1
65. Centro (Sp)	65	5.3	6.8	34.0	59.2
66. { Brandenburg (Ger)	64	2.5			
Noroeste (Sp)	64	4.3	5.8	33.3	61.0
68. Voreia Ell. (Gr)	62	3.4	21.6	30.4	47.9
69. { Sachsen (Ger)	60	4.6			
Sachsen-Ahnalt (Ger)	60	2.8			
Thurlingen (Ger)	60	2.5			
72. Sur (Sp)	58	8.2	6.8	27.0	66.3
73. { Kentriki Ell. (Gr)	57	2.6	24.5	28.7	46.8
Mecklenburg-Vor (Ger)	57	1.8			
75. Madeira (Por)	52	0.26	4.4	18.4	77.1
76. Acores (Por)	46	0.24			
UK regions					
15. Southeast England	117	17.9	0.7	20.6	78.7
30. East Anglia	100	2.1	4.9	28.6	66.5
33. Scotland	98	5.1	2.9	30.1	67.0
36. Southwest England	95	4.8	3.7	27.6	68.7
38. East Midlands	93	4.1	2.9	37.5	59.6
45. West Midlands	90	5.3	2.2	36.6	61.2
47. Northwest England	88	6.4	1.0	22.9	66.1
49. Yorks & Humberside	87	5.0	2.1	33.4	64.6
54. Northeast England	85	3.1	2.2	34.2	63.5
55. Wales	81	2.9	2.2	33.7	63.9
56. Northern Ireland	80	1.6	4.8	26.3	68.9

Source: Adapted from *Financial Times*, 23 April 1998.

the richest regions and the poorest narrowed, but that since then this convergence has ceased, and the gap between the less well-off and the most prosperous has actually increased.[16] This view is reinforced by Fagerberg and Verspagen (1996) who examine levels and growth of per-capita GDP in 70 EC regions spread across six member states. They find evidence of convergence for much of the post-war period, but of divergence in more recent years. There is, however, a dissenting voice here. Leonardi (1995) used per-capita Purchasing Power Standards (PPS) between 1970 and 1990 in nine states to show that the gap between the ten poorest regions and the ten richest regions has been to an extent reduced: from a multiple of 3 in 1970 to 2.3 in 1990. He suggests that EC redistributive policy has contributed substantially to this apparent convergence. Leonardi's findings, however, have been the subject of considerable discussion, his use of PPS being particularly controversial.

As far as studies of the specific effects of integration are concerned, there is consensus among academics (Armstrong and Harvey, 1995; Dignan, 1995; Neven and Gouyette, 1995), as well as of European institutions themselves (European Commission, 1991; European Parliament, 1991) that European integration does in fact exacerbate regional differences. Neven and Gouyette, in particular, use three different methodologies to measure convergence in output per head between 1975 and 1990. The consistent finding is that trade liberalization does increase regional differences. They suggest some catching up on the part of Southern Europe in the early 1980s, followed by at best stagnation thereafter. The regions in Northern Europe are shown to stagnate in the early 1980s, but then converge with more prosperous regions thereafter. In addition, some of the work by Fagerberg and Verspagen tends to support this view. The pure neo-liberal response to this would presumably be that markets have not been freed to a sufficient extent, and that greater liberalization would in the long run result in greater convergence. But inevitably that claim would, of course, be untestable in the foreseeable future.

Conclusion: Policy implications

It will surprise nobody that economists have differing views on the functioning of markets, and therefore on the general, and indeed the regional, impact of freeing them in the integration project. The balance of evidence suggests nonetheless some overall convergence during the early years of the EC, followed by a more recent period of divergence. There remains a broad consensus that the freeing of markets as part of the integration project has resulted in greater regional disparities in the EC/EU. This conclusion, to the extent that it can be made with confidence, has important implications, not least for the conduct of EU policy. If it is true that the very process of integration has a regionally differentiated impact, then there is a *prima-facie* case for developments in at least three areas.

First, and most obviously, this would require the existence of an effective redistributive mechanism. This is not just an issue of equity, although those arguments are worthy of serious consideration. It is also an issue of efficiency, in the context of an EU internal market in which there is limited labour mobility. That fact is unlikely to change radically in the foreseeable future given the plethora of administrative, linguistic and cultural differences that exist in contemporary Europe. There are also political issues involved, since there is a clear need for the integration project to continue to attract the enthusiasm of peripheral areas, both in the current EU and amongst potential entrants from East/Central Europe and elsewhere. As Noel Parker suggested in Chapter 1, the EU faces the classic problem of the expanding state regimes of the past: how to get, and keep, marginal territories on board. The equity and efficiency of the market to which they belong will clearly be decisive for less well-off peripheries of the EU, present and future.

At present the principal means of spatial redistribution in the EU are the structural funds, and in particular the European Regional Development Fund (ERDF). It is beyond the scope of this chapter to discuss the operation of the EU's structural funds in any detail; but, though they have certainly made an impact in individual situations, it is clear that their overall redistributive impact has been fundamentally limited by the size of the EU budget. As we have seen, the budget is very small for an organization such as the EU, particularly in view of the proposed monetary union. For the EU to have greater powers of distribution, probably including direct fiscal transfers, would require sizeably increased budgets. Yet the political difficulties here will be significant, since the transfer of fiscal powers to the supranational level would inevitably imply a major transfer of sovereignty.

Secondly, the macroeconomic policies of the new European Central Bank (ECB) and the European System of Central Banks (ESCB) in the context of EMU will have to take account of the regional impact of monetary and external exchange-rate policies, preferably in an explicit fashion. To take just one example, centrally determined interest rates that encourage a high external rate of exchange for the Euro would hinder the efforts of areas dependent on exports to areas outside of the EU, whereas the same interest rates would assist areas which import raw materials from third countries to use in manufactured goods that are destined to be sold within the EU. ECB macroeconomics ought to take account of this. Similarly, setting fiscal stances by national governments, currently to be constrained after EMU by means of the Growth and Stability Pact,[17] would need to take account of regional effects and conditions. In other words, in some circumstances scope would have to exist for national, or regional, administrations to encourage regional development, for example by easing local investment conditions even if this were to conflict with the Union's overall macroeconomic stance.

Thirdly, the very nature of EU policies towards the regions, the principles on which it is based and the way in which it is conducted, need to be reviewed. Particularly outside of the structural funds, the EU's governance in the regions is based on neo-liberal principles.[18] This may be counterproductive for some areas, insofar as freeing markets may only serve to damage relatively less prosperous and less dynamic regions. In these areas, development policies based on broader perspectives are likely to be required.

As we can see, the problem of how to foster economic growth in, and engage the loyalty of, its margins is likely to have significant knock-on effects on the EU as a whole. In the absence of a strong cultural identity or traditional coercive resources to bring less wealthy margins on board (if we leave out of consideration the push effect of fear of Russian military power and the association between NATO and EU membership), the pursuit of *economic* gains is likely to be paramount for any strategy of inclusion. But economic theory is ambivalent about the automatic benefits of joining the EU-wide market, while the political consensus, backed by the bulk of the limited empirical evidence, is unpersuaded of the gains. It is an open question how much the EU's neo-liberal assumptions, monetary principles, consultative structures or development strategies will have to change to incorporate the less developed margins. But there is clearly going to be a continuing pressure for change in the character of the EU as a whole in an effort to deal with the economic situation of its margins.

Notes

1 See Lintner (1995).
2 For example, Boyer, Aglietta, Lipiez and others in France.
3 Useful summaries of the theory of integration can be found in a number of books, including Lintner and Mazey (1991), Edye and Lintner (1996).
4 Through the effects of trade creation, trade diversion and scale economies.
5 For useful surveys of the theory of monetary integration, see, *inter alia*, Lintner (1997, 1998) and DeGrauwe (1995).
6 See Lintner (1998).
7 Currie (1997) suggests that a budget of 2 per cent of GDP might suffice for the current EMU, provided that it was entirely dedicated to redistribution and not to financing policies such as the CAP. This would permit redistribution of 20 per cent at the margin between EMU members. For a fuller treatment of fiscal federalism see Oates (1972).
8 See Viner's model in any Economics of integration text, e.g. Linter (1991).
9 The factor price equalization theorem, which is part of the basic neoclassical Hecksher–Ohlin–Samuelson theory of international trade, suggests that trade alone should lead to factor price equalization, although this cannot be definitively mathematically proved – see any international economics textbook such as Sodersten.
10 In the real world unemployment and under-employment might provide greater incentives to move.
11 In both cases there should in principle be an automatic redistribution of these gains, since migrant workers will send remittances back to the regions from where they came, and profits from capital (minus taxes) will also be remitted to the source area.
12 Myrdal (1956); see also Hirschman (1958).
13 A good survey of these can be found in Crafts (1992).
14 Perroux (1950).
15 See, for example, European Commission (1991, 1992).
16 Symes (1997).
17 Referred to as 'national rate capping' by Jonathan Michie.
18 For example, the emphasis on open markets in the Delors White Paper (European Commission, 1994).

References

Armstrong, H. and Harvey, W. (1995) 'The Regional Policy of the European Union', *Economic and Business Education*, 2(7), 4–10.

Barro, R. and Sala-i-Martin, X. (1991) 'Convergence Across States and Regions', *Brookings Papers*, 1

Barro, R. and Sala-i-Martin, X. (1992) 'Convergence', *Journal of Political Economy*, **100**, 223–51.

Borts, G. (1960) 'Equalization of Returns and Regional Economic Growth', *American Economic Review*, **50**, 235–52.

Borts, G. and Stein, J. (1964) *Economic Growth in a Free Market* (New York: Columbia University Press).

Crafts, N. (1992) 'Productivity and Growth Reconsidered', *Economic Policy*, **15**, 387–414.
Currie, D. (1997) *The Pros and Cons of EMU* (London: HM Treasury).
DeGrauwe, P. (1995) *The Economics of Monetary Integration* (Oxford: Oxford University Press).
Dignan, T. (1995) 'Regional Disparities, and Regional Policy in the European Union', *Oxford Review of Economic Policy*, **11**(2), 64–95.
Dixon, R. and Thirlwall, A. (1975) 'A Model of Regional Growth Rate Differentials along Kaldorian Lines', *Oxford Economic Papers*, **27**, 201–14.
Edye, D. and Lintner, V. (1996) *Contemporary Europe* (New York: Prentice-Hall).
European Commission (1977) *Report of the Study Group on the Role of Public Finance in European Integration (the MacDougall Report)* (Brussels).
European Commission (1991) *The Regions in the 1990s* (Brussels: Directorate General for Regional Policy).
European Commission (1994) *Growth, Competitiveness, Employment: The Challenges and Ways Forward into the Twenty-First Century* (Luxemburg: Office for Official Publications of the European Communities).
European Parliament (1991) *A New Strategy for Social and Economic Cohesion after 1992*. Regional Policy and Transport Series, No. 19, Strasbourg.
Fagerberg, J. and Verspagen, B. (1996) 'Heading for Divergence? Regional Growth in Europe Reconsidered', *Journal of Common Market Studies*, **34**(3), 431–48.
Hirschman, A. O. (1958) *The Strategy of Economic Development* (Yale: University Press).
Kaldor, N. (1970) 'The Case for Regional Policies', *Scottish Journal of Political Economy*, **17**, 337–48.
Leonardi, R. (1995) *Convergence, Cohesion and Integration in the EU* (London: Macmillan).
Lintner, V. (1995) 'The Economic Implications of Enlarging the European Union', in N. Healey (ed.), *The New European Economy* (London: Routledge), pp. 170–88.
Lintner, V. (1997) 'Monetary Integration in the EU', in V. Symes, C. Levy and J. Littlewood (eds), *The Future of Europe: Problems and Issues for the Twenty-first Century* (London: Macmillan), pp. 154–74.
Lintner, V. (1998) 'Alternative Approaches to EMU', *University of North London, European Dossier*.
Lintner, V. and Mazey, S. (1991) *The European Community: Economic and Political Aspects* (Maidenhead: McGraw-Hill).
Myrdal, G. (1956) *Economic Theory and Underdeveloped Regions* (London: Duckworth).
Neven, D. and Gouyette, C. (1995) 'Regional Convergence in the European Community', *Journal of Common Market Studies*, **33**(1), 47–66.
Oates, W. (1972) *Fiscal Federalism* (Harcourt Brace Jovanovich).
Perroux, F. (1950) 'Economic Space: Theory and Applications', *Quarterly Journal of Economics*, **64**, 89–104.
Sodersten, B., *International Economics*, various editions.
Symes, V. (1997) 'Economic and Social Convergence in Europe: a More Equal Future?', in V. Symes, C. Levy and J. Littlewood (eds), *The Future of Europe: Problems and Issues for the Twenty-First Century* (London: Macmillan), pp. 206–26.

4
Economic Development and the Periphery of the European Union

Leslie Budd

Introduction

During the past decade we have witnessed a number of profound changes in the economic organization of the European Union (EU). The important components have been the Single European Market (SEM), European Monetary Union (EMU), the single currency, and to a lesser extent the proposed expansion of the EU eastwards. The commitment of the member states of the EU to the latter is enshrined in the Treaty of Amsterdam, signed in 1997. The commitment to incorporating new member states to the east of the Oder–Niese line,[1] appears to have been the logical consequence of the momentous changes in the politics and economics of the EU's neighbours in Central and Eastern Europe. This chapter explores the relationship of the EU to its peripheries, by a discussion of the impact of the expansion of the EU. It compares the prospects of development in three different locations: the existing 'cores', the East and the Mediterranean (in particular the three Maghreb countries, Algeria, Morocco and Tunisia). The overall conclusion is that present trends in the EU are likely to create new and distinct peripheral statuses for the East and the South, while concurrently marginalizing certain presently favoured central locations. Furthermore, the prospect of economic modernization, held out by closer association with the EU, will in fact lead to a re-run of the countries of the South's colonial histories. History is likely to repeat itself as tragedy; as unfulfilled economic potential generates more political unrest.

Privileged access to the markets of some EU economic sectors has been granted to the Visegrád[2] countries, as well as Slovenia and the Baltic States. This access is seen as a precursor to full membership of the EU and, at the time of writing, eventually joining the single currency area. Despite the significant changes in the whole of Europe, little public or official attention has been granted to another important element of the EU's periphery, namely, those countries with trading access to the EU over a long period. These include the countries of the Maghreb[3] and especially Turkey which has been part of a customs union with the EU since 1962. The perennial application of Turkey for full membership of the EU has been consistently blocked. The ostensible reasons are human rights violations and little progress on settling the Cyprus dispute. There is a suspicion, often voiced, that the consistent rejection of the aspirations of the Mediterranean periphery amounts to a neo-colonial policy perspective by the EU core countries. Neo-colonialism is a situation in which countries that were subject states of former colonial empires of Europe are now subjugated by a new form of colonialism. That is, the emergence of a new international division of labour in the latter part of the twentieth century has subjected former colonies to an economic dependence on the major capitalist economies, especially in Western Europe. As a result their development is constrained and the ability to end peripheral status is severely limited. A similar argument has been developed about Japan's perspective in the Asian region (Henderson, 1998).

The profound changes in the organization of the EU suggest that issues of economic and social development and governance will be played out at the level of the region. The expansion of the EU to include the former European Free Trade Area countries of Austria, Finland and Sweden (EFTA3) and the proposed membership for certain Central and Eastern European countries suggests a strengthening of a core–periphery structure. Unless there is a stronger commitment to the inclusion of the signatories to the Barcelona Declaration (of 1995) in the EU's development, these states may become more marginal and ungovernable. A core–periphery structure is one in which the economic performance – income per head, capital formation, labour productivity, research and development, technological capacity and diffusion – differs widely between and within supranational regions, nations and/or subregions. Moreover, the degree to which social and political institutions are firmly embedded will differ within and between regions. Core regions will have more comprehensive institutions whose power and influence will be deeper and broader than in peripheral regions.

The concept of a core–periphery structure is not a simple and relative one in that, say, the European Union represents the 'core' and the Visegrád countries in Central and Eastern Europe represent the 'periphery'. Within the EU there are significant inter- and intra-regional differences. For example, Hamburg is a core region, whilst Nord–Pas–de–Calais is a peripheral one. Within these two regions there are also differences in the economic capacity and performance of localities. For policy makers, the concept of a core–periphery structure has a polemical utility in advancing their region's interests relative to other regions. For economists, it has an analytical utility in examining the development of regions from base cases. In general, the concept has metaphorical utility in assessing the complex historical development of Europe and its constituent parts from a differential perspective. This analysis draws on the approach of French theorists of centre–periphery relations in Western Europe. They suggest the notion of 'cross–cutting' regulation of centre–periphery relations to explore the complexity of relations between centre and periphery (see Meny, 1985).

What kind of convergence?

The fifteen member states of the European Union are either to be congratulated or cautioned on their present ambition. Powerful commitments to completing a single currency area (EMU), to expanding membership of the Union to include some economies in Central and Eastern Europe (Treaty of Amsterdam) and reforming regional assistance and agricultural funds (*Agenda 2000*) are creating a heady brew that may prove too intoxicating. In this environment, the commitment to creating a free trade area with the Med12 may be a step too far.

The economic and regional implications of the Maastricht Treaty have been spelt out elsewhere (Buiter, 1992; Budd, 1997). The essential problem for EMU is that the efficacy of a single currency is based on the theory of optimal currency areas (Mundell, 1961). That is, a currency area is optimal if any symmetric or asymmetric shock can be absorbed by the mobility of factors of production. A symmetric shock is one like the Gulf War in 1991 which affected all the regions and industries of the EU equally. An asymmetric shock is one like the collapse in the prices of semiconductors which affects regions and industries of the EU differently (Bayoumi and

Prasad, 1995). There are two further variants. In the first, a currency area without complete factor mobility can none the less be optimal if fiscal transfers are available to negotiate economic shocks (Eichengreen, 1990). In the second case, if there is a large amount of trade flows then a currency area can still be optimal (McKinnon, 1963). The problem with the Maastricht Treaty conditions underlying the single currency is that they are predicated on credibility theory. This theory suggests that the credibility of member states' macroeconomic policies will be established through conforming to nominal convergence criteria. These criteria relate to long-term interest rates, inflation rates, exchange-rate stability prior to the adoption of the Euro, and budget deficit and total public debt limits. The argument of EU policy-makers is that, if these targets converge, then this will create a zone of monetary stability in the single currency area. Having signalled to market participants the credibility of macroeconomic policy and its convergence to achieve price stability, real economic convergence should follow.

Looked at from another point of view, there is a lack of commitment to real economic convergence, that is, in form of output, investment, employment and growth targets. This shortcoming is likely to create political difficulties as the consequences of the single currency are played out within member states. Furthermore, the conditions for membership of the euro zone make no economic sense. The nominal convergence criteria are more about politicians' present antipathy to the public sector, shown by the commitment to constrain public expenditure by means of conforming to the fiscal criteria. Moreover, they impart a deflationary bias to EU-wide macroeconomic policy. Unless world demand picks up significantly as the new millennium begins, member states will be faced with operating austerity policies so as not to breach membership conditions. Unless the current Asian crisis is solved rapidly, the kind of 'austerity fatigue' experienced in France and Italy at the end of 1998 will spread to other member states.

New peripheries inside the EU

The lack of real economic convergence criteria and the abolition of national exchange rates means that there is no mechanism for negotiating differences in cross-border transactions within the EU. Therefore, any asymmetric regional or sectoral shocks will not be absorbed equally across the EU. The result has been summarized as:

> If real economic convergence does not occur and if economic and monetary union prevents individual regions from taking specific economic measures in order to respond effectively to shocks, there is a danger of pronounced regional differences in unemployment, incomes and growth. As long as economic shocks affect different participating countries differently and as long as sectoral and regional labour markets and/or capital mobility are low, an optimal currency area cannot be achieved. If a monetary union is nevertheless created, it will result in regional differences in unemployment, pronounced international income disparities (core–periphery effects) and hence the danger of social and political tension.
>
> (Schmidt and Straubhaar, 1995, p. 217)

It is apparent, then, that the costs of transition in the single currency area will be most keenly felt in the peripheral regions of Europe. Hence, the existing core–periphery economic structure will be reinforced by EMU. The European Commission itself has noted the potentially damaging impact of fiscal limits on the regions of the peripheral economies of Greece, Ireland, Portugal and Spain. For these economies to converge on the Maastricht criteria, deep cuts in public expenditure and large tax increases will be required. Between 1985 and 1992, government gross fixed capital formation (public investment) as a proportion of GDP declined by an average of 1 per cent in the four countries. By imposing the criteria, this investment will further decline, exacerbating the position of the poorer regions in these economies. The Commission's mordant comment will be doubly reinforced:

> [T]he recent decline in the share of public gross fixed capital formation in GDP in some of the weaker Member States suggests that there is some cause for concern . . . This is a worrying development, even for Ireland, despite the rate of economic growth being relatively high in recent years while public capital formation has been cut back. Over the longer term, such reductions are likely to depress the rate of growth.
>
> (European Commission, 1994, p. 148)

There is an argument that if these economies are to converge towards the Union average they must liberalize and attract foreign investment. But this does not stand up to close scrutiny. If internationalization

Table 4.1 Variation in regional income and unemployment by country[a]

	GDP/capita (PPS)[b]	*GDP/employee (PPS)*[b]	*Unemployment*[c]
Austria	107.0	n/a	103.9
Belgium	104.4	116.6	82.3
Denmark	106.3	84.1	102.4
Finland	95.0	n/a	112.0
France	116.6	115.7	102.7
Germany	117.9	105.2	61.6
Greece	48.1	52.8	87.5
Ireland	68.0	83.4	183.3
Italy	102.7	113.1	110.0
Luxembourg	127.2	121.9	20.1
Netherlands	101.3	98.4	79.0
Portugal	56.5	45.5	42.8
Spain	75.7	95.8	194.4
Sweden	103.0	n/a	104.7
United Kingdom	99.1	89.1	102.5

[a] Average of Level 2 regions.
[b] GDP 3-year average, 1989–91 (EU15 = 100).
[c] Unemployment 3-year average, 1991–93 (EU15 = 100).

Source: Eurostat (1994).

or globalization were to come to the rescue then one would expect a growth in foreign direct investment (FDI) in these low labour cost economies. Between 1986 and 1991 the proportion of FDI to GDP was 2.6 per cent in the four economies, the lowest being Portugal with 0.9 per cent and the highest being Spain with 7.6 per cent. The figure for the EU12 is 8.3 per cent. Despite the difficulties in measuring FDI, it is apparent that internationalization, *per se*, will not produce economic convergence of the peripheral regions. Small wonder, then, that under the 1989–93 Community Support Framework for *Objective 1* regions (those that are economically lagging), the EC Structural Funds devoted 35 per cent to investment in basic infrastructure and 22 per cent to investment in human capital (at 1994 prices) to mitigate problems of peripherality. A measure of these problems is given in Table 4.1, which shows a measure of economic disparity between the member states of the Union. This table appears to confirm the peripheral status of the economies of Greece, Ireland, Portugal and Spain, measured on a regional basis. Greece has less than 50 per cent of the EU average of GDP per capita. Although it appears to have a figure of 90 per cent of the

EU average for unemployment, this figure disguises a high degree of underemployment; that is, employment in family-related businesses as a part of informal social support. Apart from the evidence, pointed out above, about the poor growth of public capital, the four peripheral economies of the EU also have the lowest level of research and development expenditures and the lowest proportions of 16–19-year-olds in education and training programmes (see Perrons, 1992).

EU strategies to meet new core–periphery problems

The absorption of the EFTA3 into the EU seems to reinforce the core, whilst the possibility of EU membership for the Visegrád countries suggests an extended periphery. Although no official data exist for GDP per capita, on a regional basis, in the Visegrád countries, there has been a significant decline in GDP for all regions over the past five years. The benefits of liberalizing these economies and opening them up to international competition have tended to cluster new economic activities in regions around the capital cities and those bordering the EU (cf. European Commission, 1994). That is, there is better market access for regions bordering the EU and capital cities have been the biggest recipients of FDI because of good transport and communications links. This outcome will increase regional inequality within these emerging economies and thereby create political difficulties in dealing with the margins.

The evolving regional structure in Central and Eastern Europe shows that the conception and nature of core and periphery is more complex and has to include cores and peripheries within subregions and urban agglomerations. Within the core economies of the EU one finds a number of regions and subregions at the margins that have become economic shadowlands: for example, the Eastern Länder of Germany, the Mezzogiorno in Italy, parts of the north-west of England, Wales, Scotland and Northern Ireland. In France, Nord-Pas-de-Calais, bordering the Parisian Basin, is likewise counted as a peripheral region. This region has, however, a relatively wealthy core – Lille – surrounded by a number of subregional shadowlands, such as the mining basin around Mauberge. Conceiving of a simple core–periphery division to analyse economic development between the regions of the EU misses some the complex weft and weave of these territories. If crudely applied to an analysis of the peripheries of the East and South, poor policy prescription will result.

The simple conclusion that the admission of the EFTA3 will extend the core and that membership for the Visegrád states will extend the periphery cannot be drawn when one considers the consequences of the associated status granted to the latter. They have comparative cost advantages for sectors such as textiles, iron and steel and agricultural products. Between 1988 and 1992 the share of first-generation country exports to the EU rose from 3.7 per cent to 5.2 per cent, 6 per cent to 9 per cent and 3.45 per cent to 4.8 per cent for each sector respectively. The timetable for reducing EU tariffs on these exports is 1997 for iron and steel and 1998 for textiles. The long-term share of EU imports for these sectors is 4 per cent and 7 per cent respectively (cf. European Commission, 1994). The increased penetration of the industrial sectors by the Visegrád countries poses a threat to the existing EU regions, which receive *Objective 2* (declining industrial regions) Structural Funds.[4] Most of these regions are located in the core economies of the EU and not the periphery. It is wishful thinking in the extreme to imagine that the EU's limited Structural Funds programme can address industrial decline at the same rate as the reduction of tariffs imposed on some industrial sectors of the Visegrád countries takes effect. Furthermore, many regions within the EU15 will receive less regional assistance under the *Agenda 2000* programme, or none at all. This programme will reform the ERFD and the Common Agricultural Programme, with the express aim of assisting regions of Central and Eastern European economies that seek EU membership, but within a reduced total budget.

The European Commission is sanguine on the impact of the Visegrád regions on the lagging regions of the EU15. It argues that if the reform programme is successful in Central and Eastern Europe there will be net benefits to Greece and some small benefits to Spain and Portugal. However, if one looks at the distribution of outward FDI from Germany, the EU's biggest economy, the eastern rather than southern periphery has become a favoured destination, as can be seen from Table 4.2. Between 1984 and 1989 German FDI outflows accounted for 7.9 per cent of global FDI and 9.2 per cent between 1990 and 1995. In 1980 its stock of FDI was 8.4 per cent of the global total, 9.0 per cent for 1990 and 8.6 per cent for 1995 (UNCTAD, 1996). The evidence of Table 4.2 appears straightforward, but there are caveats. These include the opening up of Central and Eastern European economies after 1989, the economic nationalism of the Nordic economies and the relatively late entry of Portugal and Spain to the EU. One thing does stand out: Greece has been a member state since 1981, yet has seen little increase in

Table 4.2 German FDI outflows, 1983–94 (percentages of annual averages)

Destination	*1983–6*	*1987–90*	*1991–4*
Rest of EU	32.56	52.18	61.15
Greece	0.43	0.27	0.49
Portugal	0.31	0.71	1.22
Spain	4.89	5.36	4.40
Austria	2.21	2.71	3.19
Finland	0.11	0.14	0.18
Sweden	0.29	0.34	2.01
Central and Eastern Europe	0.12	0.55	7.60
The Maghreb	0.14	0.10	0.03

Source: OECD (various issues).

FDI from the dominant economy in the EU. It is also apparent that the growth in intra-EU FDI has been stimulated by greater economic integration, following the Single European Market (SEM) programme, with 55 per cent of the EU total going to the EU's eight biggest economies between 1991 and 1994 (Agrarwal, 1997). In 1995, FDI was the biggest source of foreign investment in Central Europe (which, for purposes of exposition, consists of the Visegrád states plus Slovenia). In dollar terms the Central European and the Baltic States received FDI inflows equivalent to 2.9 per cent of GDP. Hungary gained the highest proportion with 10.2 per cent, followed by Estonia with 5.8 per cent and the Czech Republic 5.6 per cent. Poland had a very low total of 0.9 per cent whilst the average for the Commonwealth of Independent States (CIS) was 0.7 per cent, the same figure as Russia. Clearly, the opening up of these former communist states to international economic flows and the prospect of closer association with the EU have been important factors in the growth of inward FDI.

There is an intra-EU distributional aspect to the opening up of the periphery in the East. EU countries and regions whose improved economic circumstances have been tied to inward FDI may be affected by competition for FDI. For example, one would expect the whole of Ireland (Northern Ireland and Eire) to suffer a downturn in FDI if the Visegrád countries joined the EU. However, the EC expects the impact to be minimal. The linearity of the Commission's thinking is that it treats successful reform as a homogeneous entity affecting the whole of the former Eastern Bloc. It ignores contagion effects from the difficulties of reform in Russia and its neighbouring sphere of influence in the Commonwealth of Independent States. 'Success' for the transition of the Visegrád economies is measured in terms of increased exports to the EU and the diversification of these exports to form a new international division of labour in the EU's eastern periphery. Furthermore, the Commission ignores the impact of the 'new international division of labour' on the urban hierarchy of Europe and the manner in which urban regions attempt to position themselves within this international order. Moreover, it has a conventional view of peripherality at a time when the structure of Europe's regions is suggestive of a complex of cores and peripheries.

The Visegrád countries will also have an impact on the *Objective 1* regions (lagging regions) budget. Although the budgetary position of the Visegrád countries, averaging a budget deficit of about 4 per cent between 1990 and 1993, is in the range of the EU economies their total public indebtedness is a long way from the Maastricht targets. Like many a peripheral European economy, large public debt and a non-sustainable budgetary position is a corollary of a poorly functioning tax system. FDI alone will not provide a panacea for these countries either. One of the major drawbacks of the Visegrád countries is their lack of infrastructure: a major component of regional competitiveness according to the Commission. Despite the comparative advantage of low-cost production in some industrial sectors, the Scylla of the Maastricht fiscal criteria and the Charybdis of Europe's regional hierarchy will squeeze these economies. The regional position at present is that the major cities such as Prague, Budapest and Warsaw have been recipients of the largest amount of foreign capital. The regions adjoining Western European countries have benefited also from contagion and agglomeration effects since liberalization. To grant membership to them therefore reinforces differences between the core and peripheral regions within each country. Net benefits may accrue to the dominant economies in the EU in the form of increased demand for their exports and real or portfolio capital. But the degree to which these benefits stimulate domestic demand will depend on how certain regions in the Visegrád economies can position themselves in the evolving urban and regional hierarchy of Europe. The prospect of 'trickle-down' benefits for their regions, marginalized from these hierarchies, seems to belong to the ideological and theological claims for market adjustment.

Table 4.3 General government fiscal balance and gross public debt (percentage of GDP) EFTA3: 1993, 1995 and 1997

	Fiscal balance			*Gross debt*		
	1993	*1995*	*1997*	*1993*	*1995*	*1997*
Austria	−4.2	−6.2	−3.1	57.0	69.4	73.9
Finland	−7.1	−5.6	−1.6	62.0	59.6	63.2
Sweden	−13.5	−8.1	−3.1	83.5	79.9	79.6
EU15	−5.9	−6.0	−3.4	71.3	71.2	74.3

Source: OECD (1994); Arrowsmith and Taylor (1996); Barrell *et al.* (1997).

The EFTA3 and Europe's core–periphery structure

Will the accession of the EFTA3 countries so reinforce the core as to mitigate the adverse effects of enlargement? That suggestion does not really stand up to scrutiny either. The accession of the EFTA3 into the EU in 1995 changed the core–periphery structure relatively but not absolutely. That is, its contribution in terms of growth, output, investment and income was not sufficiently above the EU average to have much distributional effect on the EU's existing periphery. Moreover, these economies have problems with their own peripheral regions. The economic impact of the accession of the EFTA3 is likely to be small. The net effect on GDP is estimated to be about 0.4 per cent. The prospect of EU membership has increased the FDI to GDP ratio in the EFTA3. This ratio declined in the mid-1980s but then rose fairly rapidly when negotiations began on the formation of the European Economic Area (EEA) and then accession to the EU, as shown in Table 4.2. The net effect appears to be that there is a positive inducement for EFTA multinationals to be located within the EU, again confirming the view that greater intra-EU FDI is associated with greater regional integration.[5] The inclusion of the EFTA3 will increase the total EU budget because they will be net contributors. This will have a positive distributional effect on the peripheral economies (cf. Baldwin, 1994). This effect will not be of great significance, however, given the reduction in the ERDF budget under the *Agenda 2000* programme and the lack of political will for re-distribution among member states, following the introduction of a single currency area.

Estimates for the fiscal position of EFTA3 show a variable performance for the different countries. Table 4.3 gives details of the fiscal reference values of Maastricht for the EFTA3 for the years 1993, 1995 and 1997. Despite the improvement in public finances, conforming to the Maastricht conditions (budget deficit maximum of 3 per cent of GDP and total public debt stock of 60 per cent of GDP) will constrain the ability of the EFTA3 to boost the EU's economic core. It is apparent that the imposition of the fiscal criteria will have a major structural impact on the budgetary position of the EFTA3 countries, and the degree to which their economies can significantly contribute to EU-wide economic prosperity. Cutbacks in public expenditure or increases in taxation will deflate their economies and reduce their ability to act as engines of growth for poorer EU economies or those on the periphery. The net impact of the EFTA3's membership is likely to be less positive than anticipated, particularly as regards distribution. If this is the case for the extending core, then it is doubly so for a shifting periphery.

CEE as a new periphery

There is likely to be a more volatile environment if the new periphery attempts to conform to stringent fiscal criteria for membership of the EU and its incipient euro zone. As noted above, the Commission believes infrastructure is essential to regional competitiveness. The decline of public direct investment in the existing periphery and their difficult fiscal position offers a salutary lesson for the new periphery. Another lesson is that international market adjustment in the form of FDI will not of itself come to the rescue.

The literature on the negative impact of regional inequality in living standards on long-run economic growth has been neglected until very recently with the appearance of new growth theory (Persson and Tabellini, 1994).[6] In the absence of regional stabilizers and real economic convergence criteria, the costs of a monetary union in Europe will be high and will bear most heavily on weaker regions (Neumann and von Hagen, 1994; Bayoumi and Eichengreen, 1995; De Grauwe and van Haverbecke, 1993). By analogy, extending the EU's periphery without resolving this problem will increase tendencies towards regional polarization, with its attendant economic and social dislocation. The Commission's sanguine position on inequality is not confirmed by evidence on general regional convergence. Appraising a neo-classical growth model, Armstrong showed that although there has been absolute convergence between 1980 and 1990, the increase in peripheral regions slowed the convergence process (Armstrong, 1994). Furthermore, there are distinctions between country-to-country and intra-country regional convergence. Armstrong concludes that core and peripheral regions may belong to different convergence 'clubs'. This would lend support to the view that the new international division of labour is generating an urban hierarchy within Europe, which is based on economic functions, and which cuts across the conventional model of core and periphery. It suggests also that the increase in peripheral regions will slow down and probably overturn the regional convergence trend. Rather than EU membership *per se*, it is the degree to which regions of the EFTA3 and Visegrád develop economic functions and activities of sufficient value and magnitude that will determine their position in the international urban hierarchy. The deepening and broadening of economic integration and the development of an optimal currency area may also be undermined, since the credibility of member states' macroeconomic policies relies on conformity to a limited set of criteria. The Treaty of Amsterdam would then take on even more the appearance of a manifesto, rather than a programme of economic and social reform for Central and Eastern Europe.

It is apparent that the member states and the Commission have embarked on a number of ambitious projects whose timing will be crucial to success. Simultaneously opening up the EU to the aspirations of its eastern periphery will present enormous difficulties and part of the cost could be denying the aspirations of the southern periphery. How then do we navigate our way out of the analytical maze to make sense of the EU's relationship to economic development in its southern periphery? The next section takes the complexity of the EU's shifting core and extending periphery further by examining the prospects for the Med12 and the Maghreb.

Euro-Mediterranean relations and the revival of a neo-colonial perspective

While the CEE countries are being reconstituted as a new periphery among others, the Mediterranean countries (which have been peripheral for centuries) find their position altering, possibly to become more disadvantageous. In this case we are looking rather at the *preservation* of the older style of colonial relationship. Like all history, it is a matter of context and conjuncture. The prospect of economic modernization offered by closer association with the EU may paradoxically reinforce a colonial relationship in the periphery to the South. In the current context the EU's orientation may condemn this new periphery to greater marginalization and increased instability. At the end of 1995 the Foreign Ministers of the EU15 and the Med12 signed a declaration at the Euro-Mediterranean Conference in Barcelona. Known henceforth as the Barcelona Declaration, it committed the participants to a programme of economic, political and social cooperation. The major commitment was to form a free trade area between the twenty-seven nations by 2010. The Declaration comprises four chapters (*Political and Security Partnership; Economic and Financial Partnership; Partnership in Social, Cultural and Human Affairs; Follow-Up*) and an annex that sets out the principal objectives and priorities of a work programme. For the purposes of exposition I focus on the second chapter, in particular the proposal for a Euro-Med Free Trade Area and the degree to which this will reinforce the peripheral status of the Mediterranean countries.

The Barcelona Declaration is the culmination of a number of initiatives to integrate the Mediterranean countries into the EU's economic and political orbit, that date back nearly thirty years. The Maghreb countries, in particular, have been at the heart of the initiatives for reasons of both history and geo-politics. Morocco and Tunisia have enjoyed free access to the EU for industrial products since 1969, under the auspices of Association Agreements, joined by Algeria in 1976 under the Cooperation Agreement for the Maghreb. These agreements also gave access to 50 per cent of agricultural products through reduced tariffs or levies. These preferences have been steadily eroded as the EU has expanded and other Association and Cooperation Agreements have been made with EU partner countries.[7] The Global Mediterranean Policy (GMP), established in the early 1970s, was replaced by the 'New Mediterranean Policy'. This policy shifted EU–Maghreb relations from an association to a partnership basis; that is, a shift to a more interactive relationship, culminating in the Barcelona Declaration.

Despite the apparent progress in EU–Mediterranean trade relations, the benefits of preferential agreements have declined. The conclusion of the Uruguay Round of the General Agreement on Tariffs and Trade (GATT) established the World Trade Organization (WTO). The Uruguay Round (*GATT94*) opened up mature markets (i.e. in the developed economies) for 'sensitive' products such as textiles, clothing and agricultural products. The sensitive products have had limited access to EU markets, despite association agreements and the 1976 agreement which extended reduced tariffs and levies for 80 per cent of agricultural products from the Maghreb. The constituent countries are caught between the rock of global

Table 4.4 Economic indicators of EU core and periphery by region

Region	*GDP growth per capita (% change)*			*GDP per capita ($000)*			*Consumer price inflation (average %)*		
	1997	*1998*	*1999*	*1997*	*1998*	*1999*	*1997*	*1998*	*1999*
EU15	2.6	2.7	2.6	21.9	21.8	23.1	1.8	1.7	1.9
EU4	2.3	2.5	2.5	23.1	23.7	24.9	1.7	1.8	1.8
Scandinavia	3.3	3.5	2.9	28.3	28.9	24.5	1.5	2.3	2.3
Visegrád	5.0	4.4	4.3	3.9	4.2	4.5	14.6	13.9	12.0
North Africa	2.0	4.1	4.3	1.18	1.17	1.21	9.3	8.5	7.3

Key: EU4 = Germany, France, UK and Italy.
Scandinavia = Denmark, Finland, Norway and Sweden.
North Africa = Algeria, Morocco, Libya, Tunisia and Sudan.

Source: Economist Intelligence Unit (1998).

liberalization of trade, under *GATT94*, and the hard place of enlarging the EU. The New Mediterranean Policy has partnership, rather than integration, as its principal aim. This is the reverse of the situation for certain Central and Eastern European countries: a point that will not go unnoticed in the capitals of the southern periphery. The relative economic position of the EU's periphery is shown in Table 4.4. It is apparent that the scale of material difference between the core and this periphery is significant. Put this together with the different institutional environment, following *GATT94*, and one begins to appreciate the magnitude of the problems. Up to now the Maghreb has benefited from the system of preferences granted by the EC and then EU. Exports grew by 12 per cent per annum between 1970 and 1993 compared to 8 per cent for Latin America (Fontagné and Péridy, 1997). Even for sensitive products there has been positive export growth. But the Maghreb's annual trade with the EU grew on average at the same rate as the whole of EU's external trade; about 11.5 per cent over the period. Given the disparity in economic performance shown in Table 4.4, a superior trade performance is needed to accelerate greater integration and its southern periphery. But the relative advantage granted by EU preferences to Maghreb exports of agricultural products, clothing and textiles, will now disappear as these sectors become integrated into the World Trade Organization (WTO).

There are four important factors in the development of the Maghreb's economic relations with the EU: financial transfers, bilateral aid, foreign direct investment and workers' remittances. Financial transfers have made little contribution to investment to growth in

GDP and are insignificant compared to workers' remittances. Under the Euro-Mediterranean policy, a number of previous protocols on financial transfers have given way to a consolidated policy approach. This approach includes a commitment to develop cross-border infrastructure and the upgrading of industries in the Maghreb and the Middle East. Over 80 per cent of bilateral aid to the Maghreb comes from France, Spain and Italy, reflecting both colonial history and migratory flows. Bilateral aid from the rest of the EU has declined from a low total between 1970 and 1992, reflecting the perception that the EU's southern 'backyard' is the concern of these three member states. Of the rest of the EU, Germany contributed the most, 2 per cent of its overseas development aid in 1992, but its total contribution declined as the significance of its eastern 'backyard' grew. The Maghreb also receives indirect bilateral aid in the form of EU member states' contributions to international aid programmes. Bilateral aid contributed more to Tunisian GDP than EU financial transfers. However, the absolute contribution to GDP was estimated to be only 0.315 per cent in 1992 (cf. Fontagné and Péridy, 1997). No estimate was made for Morocco because fixed domestic capital formation was negative in the same year. Political instability has reduced direct aid to an insignificant amount. However, the total of $1 billion direct aid from France, Italy and Spain to the Maghreb in 1991–92, and the development of debt-for-nature swaps to promote sustainable development projects in Tunisia and Morocco, suggest new cooperative relationships.

A major problem is that any stimulation of FDI will have to start from a very low base. Between 1985 and 1992 FDI accounted for only 2.5 per cent of domestic investment needs of the Maghreb. In Algeria FDI is now almost zero; though it has grown significantly in Morocco and Tunisia in the past fifteen years. There was a significant jump in 1992 with Morocco and Tunisia experiencing FDI growth of 64 per cent, with estimates of a 50 per cent growth in 1995–96. FDI is, however, limited to certain sectors. The EU accounts for half of the FDI into Morocco, with France contributing 60 per cent. The main recipients are industry, tourism, fisheries and a developing financial sector. Foreign currency inflows, rather than locally generated foreign currency debt instruments, finance 80 per cent of FDI. In the case of Tunisia over 80 per cent of FDI goes into the energy sector, the next largest recipients are tourism and building ventures with 4 per cent, followed by manufacturing and financial institutions. The majority of non-energy investment is accounted for by portfolio investment, which is sensitive to market and credit risk, i.e., the risk of a decline in general economic conditions and the possibility of firms or governments being unable to repay debts. According to EU policy-makers the stimulation of FDI flows, privatization and trade liberalization are urgent necessities. Yet the Barcelona Declaration's commitment to economic and financial partnership is limited to a wish-list of generalities. It does not commit the EU to a programme of economic and financial wherewithal which may assist these policies.

Although there has been significant growth in EU FDI in the Maghreb in the past fifteen years, it is tiny in comparison to that in other developing countries. The EU has done little in materially reducing the market risk in the Maghreb and the rest of Med12, fuelling the suspicion that it wants to direct the pace and pattern of development to its own direct advantage. The political difficulties in Algeria, and France's role in sustaining a regime of limited legitimacy, further reinforce this scepticism. At a time when the Asian crisis has not run its course, and headlines scream the possibility of global financial meltdown, a programme of economic and financial reform which relies heavily on FDI and foreign portfolio looks unlikely to deliver on its claims.

Workers' remittances are a crucial component of Morocco's and Tunisia's gross national savings and domestic absorption. In 1992 they accounted for 35 per cent and 22 per cent for the former and 7 per cent and 4 per cent for the latter, respectively. The figures are much less significant for Algeria, yielding less than 5 per cent of gross national savings and 1 per cent of domestic absorption. These flows mainly stem from the EU, with France as the largest source. Migrants from the Maghreb make up about 21 per cent of the EU's non-European migrant population. Again, France takes the majority with about 70 per cent of the Maghreb total. Despite distortions in the statistics and problems of gathering them, wage savings and transfers in the form of pensions for returned migrants provide significant relief to balance-of-payments problems. There are also a large number of transfers in kind, including motor vehicles and house purchases. There are also parallel markets in consumer goods and spare parts, where despite problems of measurement (due in the main to the official over-valuation of local currencies) there was a six-fold real per-capita increase for Morocco and a doubling for Tunisia between 1972 and 1992. The total declined for Algeria because the age cohort of migrants is younger than for its neighbours.

Hence, whereas they gain in wage savings, Algeria gains in pension transfers.

As the EU has begun to clamp down on legal migration, illegal migration will grow. As a result remittances are likely to decline. Balance-of-payments difficulties will then grow, as will the demand in the Maghreb to materially advance the New Euro-Mediterranean Policy and make the platitudes expressed in the Barcelona Declaration real. As the Barcelona Declaration includes commitments to Med12 and not just the Maghreb, the problem of 'preference competition' is likely to develop. That is, the Mediterranean countries compete with each other to gain preferential trading links. Individual EU member states will then be able to play off Med12 countries against each other to advance their particular geopolitical interests. For example, France sees Algeria and Morocco as its turf, whilst Tunisia is seeking to exploit better trade relations with Germany. Spain, on the other hand, is concerned with Moroccan claims over the Spanish Western Sahara.

There are also, of course, a number of EU-wide developments, which will impinge differentially on the Maghreb, and the rest of Med12; namely, the Single European Market (SEM) and its deepening through the single currency area, the enlargement of the EU to include the EFTA3 and Visegrád countries and *GATT94*. Despite the prospect of a larger and deeper market in the EU, the SEM has variable effects on the Maghreb. Trade creation in some key sectors is reduced by trade diversion in others. Overall, the SEM is likely to have a small negative impact, which could prove larger if protectionist sentiments evolve among EU member states, especially if the single currency project stalls. As argued above, the enlargement of the EU to include the Visegrád countries is the most worrying development for the Maghreb. The EFTA3 will not extend or shift the EU's core sufficiently, so as to stimulate 'trickle-down' development in the periphery and margins of Europe. On the other hand, by the end of 1998 the Visegrád countries will face complete progressive dismantling of tariff barriers for sensitive products. The impact of Visegrád membership of the EU will then be similar to the earlier accession of Greece, Portugal and Spain: the Maghreb will experience a loss of preferential access to the EU's markets. This disadvantage may be balanced by the expansion of the Maghreb's markets for some sectors such as agricultural products. However, trade and investment diversion is a consequence of incorporating the eastern periphery, with its higher incomes and greater FDI and portfolio investment possibilities. Furthermore, the Visegrád countries, unlike the Maghreb, do not have a colonial history with EU member states, which leaves them with a greater degree of discretion in relation to the international economy.

Even without greater integration of Central and Eastern European economies, *GATT94* poses great difficulties for the Maghreb and the rest of the Med12. New sets of rules for international trade, established by the Uruguay Round of GATT particularly as regards food and textiles, are expected to provide the basis for global growth. But, for the Maghreb, *GATT94* will further erode its preferential relationship with the EU. It is estimated that manufacturing exports to the EU will decline by 5.2 per cent for Morocco and 27.2 per cent for Tunisia, over a ten-year period (Fontagné and Péridy, 1995).

Perhaps, more crucially the free trade area to be set up between the EU and the Med12 by 2010 does not hold out significant advantages for the Maghreb. Like *GATT94* it will reduce existing preferences that the Maghreb enjoys with the EU. Moreover, a 'hub and spoke' structure is likely to develop among the Med12 (Fontagné and Péridy, 1995) with the EU treating 'hub' Med12 countries differently from 'spoke' countries.[8] EU policy will privilege some countries in the southern periphery against the interests of others. Furthermore, the Barcelona Declaration is concerned with EU–Med12 trade relations and not intra-Med12 trade relations. Given the different levels of bilateral aid to the EU and the Maghreb, those EU member states with closer ties will be able to exploit their geopolitical position. In essence this adds up to conditions for the relationships of colonial history to be replicated under the rubric of global liberalization. In other words, the new neo-colonial powers of the EU suggest that the southern periphery cannot buck the international markets but can only negotiate the markets through their neo-colonial relationships. Without greater cooperation and integration between the Med12, materially underwritten by the EU, this unhappy scenario will continue. In consequence, without a real commitment by the EU to a comprehensive programme of economic development that is sustainable for the whole of the Med12 area, some of the recent political instability and violence will get worse. The perception of a neo-colonial agenda behind a programme of liberal economic reform may lead the southern periphery to reject economic modernization and retreat into more fundamentalist culture and belief. The unhappy historical experience of France and Algeria, played out today in the *banlieues* and *quartiers* of France's major cities and the hospitals and morgues in Algeria appears to bear this out. It is not an enticing prospect for the EU to deal with. It is one to be addressed, however, if the EU policy of developing 'a Europe of the regions' is not to give way to 'a Europe of the margins', whose consequences would impact not only in EU's hinterland.

Conclusion

The EU has chosen a less than propitious time to undertake a number of ambitious and simultaneous projects. The single currency area to complete the SEM, the enlargement of the EU and concluding a free trade agreement with its southern periphery come together at a time when the prospects for the world economy are not encouraging. The conditions for establishing the euro zone are deflationary. The enlargement of the EU to include the EFTA3 countries will not sufficiently boost the EU's economic core to compensate for the extension of the periphery to include the Visegrád countries. The EFTA3 countries have had their growth prospects constrained, moreover, by conforming to nominal convergence criteria.

The EU has recognized the efficacy of incorporating the southern periphery into its ambit. However, the platitudes of the Barcelona Declaration do little to dispel the suspicion that an ideology of the imperatives of globalization and liberalization is being used to establish a neo-colonial relationship. EU policy in general appears to be strongly influenced by the geopolitical strategies of the former colonial powers of the South, in particular France and Spain. These EU member states seem to want any programme of development in the South to be subordinate to their national interests which regard the Maghreb countries as part of their own backyard. The evidence about the Maghreb, advanced here, has done little to dispel this suspicion. Without significant commitment of the resources needed to generate broader and deeper regional development in the growing periphery of the EU, negative feedback effects will be generated in the EU's regional core which may not be manageable: for example, problems and political unrest among emigrant populations and the rise of nationalist and racist sentiment within the EU. In the absence of this material commitment the margins of Europe will arrive at all our doorsteps and be no longer governable. The populations of the southern periphery will see the prospect of economic development and modernization as a chimera that strengthens rather than weakens their colonial status. In the case of the periphery in the East, there is a danger that these countries replicate the Asian crisis. They may fix their currencies to the euro, fund FDI by short-term debt and generate large balance-of-payments problems. The fall-out from this kind of scenario would be similar to that currently experienced in Russia. The possibility of ungovernable margins of Europe is something that EU policy-makers should address urgently.

Notes

1 The Oder–Niese line was the natural barrier which came to divide Western Europe physically and symbolically from the Communist bloc of Eastern Europe.
2 The Visegrád countries (the Czech Republic, Hungary, Poland and Slovakia) were granted access to the EU markets in steel, textiles and footwear.
3 The Maghreb consists of Algeria, Libya, Mauritania, Morocco, Tunisia and the Western Sahara, of which Algeria, Morocco and Tunisia are signatories to the Barcelona Declaration. Again for purposes of exposition these three countries are collectively termed 'the Maghreb'.
4 The Structural Funds are part of the European Regional Development Fund (ERDF) designed to give regional assistance to regions in economic transition. The major components are **Objective 1**, lagging regions: **Objective 2**, declining industrial regions; **Objective 5a**, rural areas.
5 The Single European Market has opened up new market opportunities because of a larger market with lower transaction costs. There is evidence to suggest that those European economies which stay outside the EU face the threat of their largest corporate companies moving operations to the EU to gain better market access for their products and services (Baldwin, 1994).
6 New growth has drawn on German literature associated with the Social Market Economy of the 1950s, albeit unwittingly. New growth theory does not refer to the experience of immediate post-war Germany; its proponents appear ignorant of this experience (cf. Schmidt and Straubhaar, 1995).
7 These included Malta, Spain and Israel in 1970. Cyprus, Lebanon, Egypt and Portugal joined them in 1972. The Global Mediterranean Policy (GMP) was created soon after to harmonize all external policies. Under the GMP, the Maghreb Co-operation Agreement was concluded in 1976, followed by the Mashrek countries of Egypt, Jordan, Lebanon and Syria in 1977 and Yugoslavia in 1980.
8 This phrase comes from US aviation policy, whereby airlines establish hub airports from which their services flow along spokes to other destinations.

References

Agrarwal, J. (1997) 'European Integration and German FDI: Implications for Domestic Investment and Central European Economies', *National Institute Economic Review*, no. 2, pp. 100–11.

Armstrong, H. (1994) 'Convergence versus Divergence in the European Union Regional Growth Process 1950–90'. Management School, Lancaster University, Discussion Paper EC19/94.

Arrowsmith, J. and Taylor, C. (1996) *Unresolved Issues on the Way to a Single Currency* (London: National Institute of Economic and Social Research).

Baldwin, R. (1994) 'The EFTA Enlargement of the European Union'. Centre for Economic Policy Research Occasional Paper no. 16.

Barrell, R., Morgan, J., Pain, N. and Hubert, F. (1997) 'The World Economy', *National Institute Economic Review*, pp. 25–55.

Bayoumi, T. and Eichengreen, B. (1995) *Restraining Yourself: The Implications of Fiscal Rules for Economic Stabilization.* IMF Staff Paper 42, pp. 32–48.

Bayoumi, T. and Prasad, C. (1995) 'Currency Unions, Economic Fluctuations: Some Empirical Evidence'. Centre for Economic Policy Discussion Paper no. 1172, May.

Budd, L. (1997) 'Regional Integration and Convergence and the Problems of Fiscal and Monetary Systems: Some Lessons for Eastern Europe', *Regional Studies*, **31**(6), 559–70.

Buiter, W. (1992) 'Should We Worry about the Fiscal Numerology of Maastricht?' Economic Growth Center, Yale University, Discussion Paper no. 654.

De Grauwe, P. and van Haverbecke, W. (1993) 'Is Europe an Optimum Currency Area?', in P. R. Masson and M. P. Taylor (eds), *Policy Issues in the Operation of Currency Unions* (Cambridge: Cambridge University Press), pp. 111–29.

Economist Intelligence Unit (1998) *Regional Data* (London: Economist Intelligence Unit).

Eichengreen, B. (1990) 'One Money for Europe? Lessons from the US Currency Union', *Economic Policy*, no. 10, April, pp. 117–87.

European Commission (1994) *Competitiveness and Cohesion: Trends in the Regions*. Fifth Periodic Report on the Social and Economic Situation and Development of the Regions of the Community (Luxembourg: Office for Official Publications of the European Commission).

Eurostat (1994) *Rapid Reports: Regions*, no. 2 (Luxembourg: Eurostat).

Fontagné, L. and Péridy (1995) 'Uruguay Round et PVD: le cas de l'Afrique du Nord', *Revue Economique*, **46**(3) (May), 63–81.

Fontagné, L. and Péridy (1997) *The EU and the Maghreb* (Paris: Organization for Economic Co-operation and Development).

Henderson, J. (1998) *Globalisation of High Technology Production* (London: Routledge).

McKinnon, R. (1963) 'Optimal Currency Area', *American Economic Review*, **53** (September), 717–25.

Meny, Y. (1985) 'Cross-cutting Regulation in Centre–Periphery Relations', in Y. Meny and V. Wright (eds), *Centre–Periphery Relations in Western Europe* (London: Routledge, Kegan Paul).

Mundell, R. (1961) 'A Theory of Optimal Currency Areas', *American Economic Review*, **51** (September), 657–65.

Neumann, M. and von Hagen, J. (1994) 'How Far Away is EMU?', *Review of Economics and Statistics*, **76**(2), 236–44.
Organization for Economic Co-operation and Development (various issues) *International Direct Investment Statistics Yearbook* (Paris: OECD).
Organization for Economic Co-operation and Development (1994) *Economic Outlook*, December (Paris: OECD).
Pain, N. and Lansbury, M. (1997) 'Regional Economic Integration and Foreign Direct Investment in Europe', *National Institute Review*, no. 2, 87–99.
Perrons, D. (1992) 'The Regions and the Single Market', in M. Dunford and G. Kafkalas (eds), *Cities and Regions in the New Europe* (London: Bellhaven Press), pp. 56–73.
Persson, T. and Tabellini, G. (1994) 'Is Inequality Harmful for Growth?', *American Economic Review*, **84**, 600–21.
Schmidt, C. and Straubhaar, F. (1995) 'Maastricht II: Are Real Convergence Criteria Needed?', *Intereconomics*, September/October, pp. 211–20.
UNCTAD (1996) *Trade and Development Report* (New York: United Nations).

Part II

Managing Economic Forces at the Margins

5
EU Trade and Aid Strategy in the Southern and Eastern Mediterranean

Mina Toksoz

The EU's strategies for the economic and political management of its Eastern and Southern Mediterranean margins have changed over the years. This chapter examines the evolution of those strategies, and reactions to them in the different marginal spaces of the eastern and southern Mediterranean (ESM) countries. Even ignoring EU–Balkan relations (which would clearly require a further chapter to themselves) the economic, political, cultural and historical patterns across this southeastern margin of the EU reveal diverse, overlapping dynamics. Powerful centrifugal and centripetal forces have simultaneously brought about internal economic and political fragmentation while promoting an unstable tendency towards a limited form of economic integration with the EU itself. The EU has a struggle to manage this complex margin on its southern edge.

The changing EU strategy

In the 1960s the Mediterranean basin was seen as an integral part of the European economic space, and as part of the European single market. Commission statements confirmed that 'the development of the Mediterranean basin [was] a natural extension of the European integration' (European Commission, July 1971). These objectives were expressed in a series of Association Agreements with the Mediterranean basin countries. However, unlike in Latin America (where the USA is the undisputed leader), or in Southeast Asia (where Japan provided the motor of economic development), in the Mediterranean the EU has failed to provide either a coherent political leadership or sufficient economic dynamic for growth. By the end of the 1970s, security concerns (such as immigration or terrorism) and the threat political Islam posed to the prevailing order in the region had come to the fore in EU pronouncements. But beyond them, the EU's attempts to form a coherent policy towards the region were thwarted by the north–south divisions *within* the Union, which made community-wide initiatives difficult. From the late 1980s the collapse of the Soviet Union shifted the interests of the northern EU countries (led by Germany) towards the east. Economic, cultural, and political distances grew, separating the ESM countries further from the EU as a whole. The southern EU states, for their part, began to search for a new relationship with their neighbours on the southern shores of the Mediterranean.

The countries on the Mediterranean margins of the EU are a heterogeneous and politically divided group. One of the problems with EU policies towards the region is that this fact is not sufficiently recognized. The patterns of ESM countries' relations both to one another and to different political and economic 'cores' are complex and their levels of integration with the EU vary in nature and degree.

- Cyprus began its membership negotiations with the EU in 1998 but is in dispute with Turkey over the status of the Turkish Cypriot enclave in northern Cyprus. Malta had decided to opt out of the membership track but has more recently decided to get in the queue once again.
- Turkey has had a customs union agreement with the EU since 1996 which makes it part of the EU single market. But Turkey's application for EU membership was turned down, as was Morocco's in the late 1980s. Political relations between Turkey and the EU deteriorated almost to breaking point in 1998, after Turkey was excluded from the then list of potential members in the next stage of EU enlargement.
- So long as Israel remains isolated from its Arab neighbours over the Palestinian question, Israel combines very close trade and economic ties to the EU with tense political relations.
- Thus, both Turkey and Israel, the two biggest economies in the region with the greatest degree of *economic* integration with the EU, remain the most wary of the EU *politically*, and look instead to the USA for leadership. The USA itself remains a major, separate player both politically and economically. It is, for example, the biggest investor in Egypt.

Table 5.1 GDP growth (average per year)

	1985–6	*1986–90*	*1991–5*	*1996*	*1997*	*1998*
Algeria	4.6	0.3	0.8	3.43	2.5	3.8
Egypt	7.3	3.9	3.4	4.9	5.3	4.9
Israel	3.6	4.2	6.1	4.5	1.9	0.9
Jordan	7.6	−0.8	7.6	1.0	2.8	1.6
Lebanon	n/a	n/a	12.8	4.0	3.5	3.2
Libya	−3.4	−0.7	−0.9	1.5	0.5	−1.0
Morocco	4.3	4.5	1.4	11.5	−2.2	6.1
Syria	2.5	1.8	7.4	2.2	1.6	0.8
Tunisia	4.8	3.0	3.9	7.0	5.4	5.2
Turkey (GNP)	3.7	5.6	3.3	7.4	8.0	4.5

Source: Economist Intelligence Unit.

- Even though it is one of the Union's main energy suppliers – and not long ago was a part of France – Algeria's civil war has put it politically at odds with the EU.
- Under a US-led global consensus, Libya and Syria have been isolated by the EU because of their support for terrorist groups active in the region. However, an initiative in July 1998 resulted (much to the consternation of the USA) in a new agreement between Italy and Libya which aims to bring Libya into the Euro-Med dialogue.

Internal economic divergences in the ESM: oil, industrialization, protectionism

Heterogeneity is also shown in the relative lack of integration *between* the Mediterranean economies (Toksoz, 1993). Whereas EU strategies attempt to treat ESM countries as a more or less homogeneous periphery, in reality there exist within the region itself patterns of core and periphery and of overlapping margins which are relatively independent of any development *vis-à-vis* the EU.

A semblance of a regional economic division of labour had emerged in the boom years of the late 1960s and the 1970s when Lebanon became the financial and services centre for the oil exporters in the region. GDP growth in the Middle East and North African (MENA) economies was then on average amongst the highest in the world, at around 6.7 per cent per year. The huge construction projects in the Gulf and Libya brought large movements of labour across national boundaries within the region, and a flourishing trade in construction services. The oil surpluses of the 1970s led to the establishment of a number of pan-Arab financial institutions, such as the Arab Fund for Economic and Social Development and the Arab Monetary Fund. Support for Arab political and economic integration grew. But the collapse of the Lebanese economy from the mid-1970s signalled the return to economic segmentation, accompanied by parallel relations with economies *outside* the region altogether. In the 1990s only around 7–8 per cent of total trade by the MENA countries has been with each other; while some 40 per cent of trade is with the EU. Growth has also diverged (Table 5.1). In the second half of the 1980s those economies heavily dependent on oil and other commodity exports (Algeria, Egypt, Libya, Syria) suffered. Countries such as Israel and Turkey, which had managed to switch the bulk of their exports to manufacturing, grew steadily.[1]

In contrast to global trends in economic policies in the 1980s, which were based on liberal, private investment and export-led strategies, much of the MENA continued with the old model of import substitution, and protectionist, state-led industrialization. Thus by 1990, while effective tariff rates on imports for the OECD had fallen to 2.8 per cent, countries in the southern Mediterranean still had some of the highest protectionist walls in the world. Both Morocco's and Tunisia's rates were over 19 per cent, compared with the average even for the non-OECD of 14.6 per cent (Abed, 1998, p.18). The decline in oil prices from the mid-1980s reinforced the relative inward orientation of these economies with the share of exports of goods and services to GDP declining for many (again there were exceptions to this trend such as Israel, Morocco, and Turkey). Government budgets in the region still relied heavily on revenues from import duties. As a share of tax revenues these had fallen to 2.57 per cent in the OECD, while in the Middle Eastern countries that figure was still at 26.98 per cent, second only to that for African economies.

Yet again there was no common pattern across the Mediterranean rim as a whole. Both Turkey and Israel had lowered their trade barriers to OECD levels. Except for agriculture and a few industrial sectors, Turkey eliminated all tariffs on EU imports in 1996. Thus, although by 1995 average (unweighted) effective tariffs for the southern Mediterranean countries were still a high 15.31 per cent, in Jordan this had fallen to 12.3 per cent and in Syria to 9.9 per cent. Trade liberalization was accompanied by structural reform, which encouraged more private investment and policies to attract foreign investors. Huge strides were made in recent years in Jordan,

Table 5.2 Ratio of export of goods and services to GDP

	1980	*1985*	*1990*	*1995*	*1997*
Algeria	34.3	23.5	24.3	27.5	29.6
Egypt	27.9	17.7	20.2	22.0	20.6
Israel	43.8	44.7	34.7	31.7	31.2
Jordan	38.2	38.7	61.9	52.4	49.9
Libya	60.1	44.7	40.5	29.1	43.0
Morocco	15.0	19.9	24.6	26.9	26.7
Syria	18.2	12.0	28.3	32.1	32.8
Tunisia	40.6	32.6	43.6	44.7	42.2
Turkey	5.2	16.4	13.8	21.3	26.0

Source: Economist Intelligence Unit.

Tunisia, Morocco and Egypt. Privatization programmes and the development of capital markets have begun to attract investment and carry the region out of virtual stagnation seen during the decade between 1985 and 1995.

In sum, since the 1970s the ESM countries have diverged from each other, according to a complex of factors: relative degree of dependence on oil; the weight of the state sector and/or its degree of dependence on tariff income; choices made by political-economic elites to follow, or not, the global trend towards liberalization; foreign direct investment flows from Europe and beyond; and discriminations produced by US-led global power politics (such as the integration of Israel or the isolation of Libya). All this has left behind earlier EU trade-management strategies, which relied upon a benign economic impact from association with the EU and presupposed that the ESMs' territorial contiguity would of itself foster internal economic integration amongst them. The Mediterranean margin was subject to a wider range of centrifugal forces than EU economic strategies could envisage.

Barcelona and the Euro-Med agreements

By 1992 EU–Mediterranean relations had almost collapsed. Then came progress in the Middle East peace process. The 1993 Oslo accord offered a greater political role for the EU in the region, and set a new optimistic tone for Middle East–EU relations. The crisis in the Balkans had shown the limitations of the EU's foreign policy machinery and greater efforts were now being made to establish institutions to forge a common foreign and security policy (the

CFSP), especially in relation to the Mediterranean. The Spanish presidency of the European Council, in the second half of 1995, led to the launch in November of the Barcelona Agreement which tried to put EU–Mediterranean relations on a new footing.

Although the political atmosphere in the region had changed, the proposals in the 1995 agreement had in fact been tried before. The precursor to Barcelona had been the 1987 Euro-Maghreb partnership, which had been drawn up by the EU commissioner responsible for the region at that time, Manuel Marin. This partnership had completely fizzled out, killed by disputes between Spain and Morocco over tomatoes and fishing rights and by the start of the civil war in Algeria. However its main themes were all repackaged in the Barcelona Agreement: the shift from aid to loans, the focus on helping private-sector small business, establishment of a Euro-Maghreb Development Bank, the liberalization of trade accompanied by balance-of-payments support for countries undergoing trade liberalization.

But the Barcelona Agreement also went further, by stipulating the formation of a Mediterranean Free Trade area by 2012. To this end new Association Agreements were proposed which would replace the old Co-operation Agreements from the 1970s. These agreements, which were to be implemented over a 12-year period, were essentially the same as those proposed to the Central and East European countries, but did not, except in the case of Turkey, envisage a process leading to full membership. After much wrangling between the northern and southern EU states, a financial package was agreed which, when all the components were added together, came to around ECU10bn over 1995–99. This was to be spread between eleven states: Algeria, Cyprus, Egypt, Israel, Jordan, Lebanon, Malta, Morocco, Syria, Turkey, Tunisia. The financial support was to help compensate for the loss in revenues resulting from the tariff reductions and to finance infrastructural projects. Finally, the agreement also sought to increase EU–Mediterranean political dialogue and to establish a charter, modelled on the East European one, for peace and stability in the region.

Yet Southern Mediterranean countries (SMCs) saw few gains in the new Association Agreements, which aimed at liberalizing trade in industrial goods, but excluded services, agriculture and, of course – given the European anxieties over immigration – the free circulation of labour. The SMCs already had free access to the EU markets in the industrial sectors. They were not sure that, if they reduced their tariff barriers, their domestic manufacturing industries could survive against competition from EU products. Tunisia and Morocco, for example, reckon that 30 per cent and 60 per cent of their industrial sector could be wiped out by EU competition (Joffe, 1998). The SMCs insisted that they needed long transition periods and more financial support. The promise of increased FDI flows once they restructured and reformed their economies seemed too uncertain. On the other hand, the bulk of exports to the EU from Morocco, Tunisia, and Egypt are agricultural. So they sought more access for agricultural exports, where they had a competitive advantage over EU goods. But negotiations on that have been postponed until 2001. In short, for the economies of Mediterranean North Africa, the exclusion of agricultural goods from the liberalization programme limited the potential gains of the Association Agreements from the start (Galal and Hoekman, 1997).

One further problem with the trade agreements with the EU is that they are doing little to overcome the segmentation of the Mediterranean markets. One of the aims of the Euro-Med agreement was to encourage greater integration of the regional economies, not just with the EU but with each other. Yet, paradoxically, the 'rules of origin' clauses (which allow imports into the EU according to a specified ratio of domestic content) have tended to reinforce the existing patterns of trade and discourage horizontal trade among the SMCs. Rather than integrating, the trade agreements reinforce the divergencies between ESM countries.

The deterioration of the regional political atmosphere after the election of Binyamin Netanyahu as prime minister in Israel in June 1996 provided a gloomy background to these fraught trade negotiations. The Euro-Med meeting in Malta in 1997, which many high-level EU ministers did not even bother to attend, was dominated by debates over the peace process. The Arab states demanded that the EU freeze all relations with Israel. The release of the funds from the financial protocol was also held up by Greece, which objected to the allocation of EU funds to Turkey. The 1998 meeting in Palermo, Sicily, was equally fraught, with very little progress made; by August Egypt was threatening to suspend trade negotiations with the EU.

Yet, despite these problems and reservations, four states have so far signed the Association Agreements: Jordan, Morocco, Tunisia and Israel, together with the Palestine Authority (although this remains blocked by Israeli objections). Another three agreements are currently being negotiated: with Algeria, Egypt and Syria. With the forthcoming eastward expansion of the EU, the SMCs realized they had to reinforce their links with their biggest market in order to maintain their market shares. A free trade area by 2012 is not impossible.

These economies have also moved on, not because of the Euro-Med agreements but more because the political elites have reacted to present global conditions with new liberalizing policies to ensure economic growth. In 1996 the IMF was warning that the Mediterranean North Africa's growth in trade had lagged behind that of almost all other regions of the world and that real per-capita incomes had declined dangerously over the previous decade. But by early 1998 an estimate of GDP growth of around 5–6 per cent was not looking unrealistic for the most reforming economies, such as Egypt. Although FDI flows were still only trickling in, they had picked up in response to the privatizations in Egypt, Morocco and Tunisia.

The high expectations regarding growth in the first half of 1998 will probably have to be revised down in the light of the global crisis in equity markets and the slowdown in world trade growth following the crisis in Asia. The reform and restructuring programmes of the southern Mediterranean rim economies will be that much more difficult in the global economic environment we see for the next few years. However, one advantage these economies have had is that, because they never had much portfolio investments, there is little negative fall-out from the contagion which is hitting many emerging market economies. Indeed, in September, while stock markets all around the world collapsed, in North Africa the markets continued on their own dynamic, showing little correlation with the global trends.

Structural consequences for the politics of ESM countries

The liberalization of the ESM has significant political consequences. Shrinking the state's role in the economy through the privatization of state enterprises loosens the client relations which have maintained the political *status quo*. It also undermines the role as a form of social welfare which, given the general absence of unemployment benefits in these economies, has been played by over-manned state enterprises. Moreover, as pointed out above, trade liberalization exposes domestic manufacturing to competition from EU imports which it cannot withstand. The resulting increase in economic insecurity makes public opinion more receptive to nationalist, protectionist, anti-Western political themes – the most coherent exponents of which are the Islamic parties. A new entrepreneurial, consumer-oriented middle class does eventually emerge, as can be seen in Turkey. However, the process fragments society sharply and is fraught with dangers for the pro-Western political elite, who find support for their liberalizing case undermined.

Thus, while some progress has been achieved in closer economic integration, it has not had much positive effect on tensions in EU–Mediterranean political and cultural relations. One locus of the divide, which has come increasingly to the fore in EU–Mediterranean dialogue over the past decade, is religion. Muslims' view of the EU as a Christian club was reinforced over 1997–98 with the exclusion of Turkey from the list of prospective members for the next stage of enlargement. A conference of EU Christian Democrat Parties, attended by the then chancellor of Germany, Helmut Kohl, openly stated that Christianity was an essential cultural glue for the EU project – apparently oblivious of the large Muslim communities already living within EU member states.

Reinforcing these divisions is the problem of extricating relationships from the colonial legacy. A recent example of the difficulties of this was the agreement between Italy and Libya, which took 13 years of discussions. It required the Italian government to apologize for the 'suffering caused to the people of Libya by colonization' – some 100,000 Libyans, around 12 per cent of the inter-war population, were killed during the Italian occupation of 1911–45 (Puccioni, 1998). Given Libya's desperate need for international allies, Italy was not asked to pay any compensation. However, it is unlikely that France, Germany or Britain could be let off so lightly. Conversely, political opinion in the EU countries is not inclined to be at all sensitive to anticolonial sentiments on the southern shores of the Mediterranean. Rather, since the Algerian crisis, political opinion in the EU has linked political instability in the Maghreb with threats of mass immigration into continental Europe. This trend is increasingly being picked up by centre parties, as seen in the recent election campaign by CDU/CSU in Germany.

A recent US Rand corporation study (Lesser, 1996) described the Mediterranean as 'among the world's most dramatic demographic, economic and cultural fault-lines'. As discussed above, the changing economic conditions which bring about closer integration of these economies with the EU are simultaneously wrenching the ESM countries *away* politically by feeding anti-Western sentiments in the Southern and Eastern Mediterranean. Even leaving aside difficult political issues such as the Palestinian question and the north Cyprus situation, the mutual suspicion with which both sides of the Mediterranean regard each other goes deep and it will take substantially more economic development and progress south of the Mediterranean before relations improve significantly.

In sum, Europe's efforts to manage its relationship with the Mediterranean countries on its southern margin are caught between contrary effects and conflicting political impulses. Earlier expectations of benign, mutually beneficial relations, based on the straightforward benefits of economic growth within the orbit of the EU, have not been borne out. Whilst mutually agreed economic development remains the principal focus of attention, there is scepticism on the ESM side about the intentions of the EU. Within Europe itself, an obsession with migration has fed demands for Europe to be primarily a fortress against migration from the ESM countries. Meanwhile larger economic forces have brought about economic fragmentation between the countries in the region, while liberalizing responses have become associated with economic insecurity that undermines the pro-Europe policies of local political elites.

Note

1 Although Jordan does not produce oil, its function as an entrepôt economy renders it vulnerable to regional fluctuations in growth.

References

Abed, G. T. (1998) *Trade Liberalisation and Tax Reform in the Southern Mediterranean Region* (Washington, DC: IMF Working Paper).

European Commission (1971) *Community Development Cooperation Policy* (July).

Galal, A. and Hoekman, B. (eds) (1997) *Regional Partners in Global Markets: Limits and Possibilities of the Euro-Med Agreements* (London: Centre for Economic and Policy Research – CEPR).

Joffe, G. (1998) *The Euro-Mediterranean Partnership: Two Years after Barcelona.* RIIA Briefing Paper (London: Royal Institute of International Affairs).

Lesser, I. O. (1996) *Southern Europe and the Maghreb: US Interests and Policy Perspectives* (Santa Monica, CA: Rand Corporation).

Puccioni, M. (1998) 'Rome Gives Pariah a Foothold in Europe', *The European*, 27 July.

Toksoz, M. (1993) *Pockets of Influence: The EC's Faltering Impact on the Arab Economies* (London: Royal Institute of International Affairs).

6
Multinational Investment on the Periphery of the European Union

Christopher Flockton

Multinational enterprises, trade and investment at the margins of the EU

Spurred on by the progressive liberalization of capital movements through the 1980s, in recent decades foreign direct investment (FDI) by multinational corporations has played a major role in global integration of the world market. In consequence, a significant proportion of world trade in manufactures is now thought to take the form of intra-firm trade; namely, shipments between subsidiaries of a single company. Multinational investment can exert a profound influence therefore in integrating further the separate economies of the EU. Moreover, the mobility of capital to less-developed regions is thought to be critical in reducing disparities in output, productivity and incomes. Thus, FDI is considered essential in assisting Central and East European countries (CEE) in their transition to the market and in tying them closely to the European Union.

We may conceive an integrating Europe as a core, the single market of the current fifteen members, with concentric circles of preferential trading zones radiating outwards, each with diminishing intensity, from it. Thus, the European Economic Area, the Association Agreements (chiefly with Mediterranean countries), the Europe Agreements with Central and Eastern Europe, and the Lomé Agreements with Associated Caribbean and Pacific countries are all zones of differential access to the single market, often with asymmetrical trade agreements governing trade flows in both directions between the EU and the partner. A multinational corporation thinking of investing in a partner country so as to gain access to the EU single market will consider the degree of preferential access for manufactured products. In this environment, FDI, in parallel with trade links may then exert a key influence in binding economic regions together. So, the question naturally arises whether according preference to a new economic region – in this case Central and Eastern Europe – will merely lead to a 'diversion' of FDI from some existing recipients, such as southern Europe (Baldwin, 1994), thus redistributing effects within the trading zones outlined above. This is the focus of the present enquiry.

Multinational investment is a major factor in development and trade, though its influence is not intrinsically benign. Theoretically, the foreign investment can help in development by covering the gap between investment needs and the low savings levels in less-developed countries and regions. Likewise, it can promote the transfer of technology or advanced managerial skills and, through capital mobility, equilibrate the balance of payments. Multinational investment is also a fundamental factor in the world division of labour: by the development of countries' comparative advantage it can foster the re-allocation of production world-wide. In so doing it promotes growth and strongly influences goods flows among regions of the world. Of course, these theoretical gains can be subverted by multinationals when they exercise market power, pursuing monopolistic or oligopolistic strategies – when, for example, they engage in so-called transfer pricing. In this fashion they can undermine or evade the policy choices of national governments. In less-developed regions the dependence of the regional labour market on a large foreign employer can equally lead to exploitative relationships. However, the view taken here is that in net terms the gains of multinational investment are substantially positive: it has a key role to play in both the further integration of the present fifteen members of the EU, and in the progressive integration of central European countries into this West European grouping.

In practice the greater part of the stock of world FDI is held in developed countries, in the so-called 'Triad' of Japan, the USA and the EU: the much smaller share held by less-developed countries has been declining since the 1970s. This implies that FDI is conducted primarily among countries of similar levels of development, and therefore that the search for low-cost production sites (primarily low-wage regions) is far from the dominant motive. Rather, penetration of existing markets, export servicing, the strategic development of new product markets, the acquisition of high technology, and overcoming tariff and non-tariff barriers are all motives for FDI, not to mention the refined oligopolistic strategies played among a group of key multinationals each seeking a global presence (Caves, 1996; Dunning, 1993). There is, of course, a close linkage between FDI flows and trade flows, and the question arises whether FDI substitutes for exports, or whether it fosters them. This is a question with many levels of argument, but the prevalent view is that, on balance, FDI promotes exports from both the investing and host countries, although empirically there will be considerable variation.

We can envisage a number of different scenarios. A multinational which previously exported from the home country may perhaps establish a manufacturing plant in the host country, which then serves that market from the new, local base. If the host country has a comparative advantage in the production of the good, be it a low labour cost or raw material-based advantage, then the country will develop an export potential. This 'comparative advantage'-based trade is known as *inter*-industry trade: where countries exchange dissimilar products. Alternatively, the multinational corporation may establish a subsidiary which either manufactures items from within the corporation's final product range, or which manufactures components or acts as an assembler of components. In the first case the multinational may exchange its finished products through trade between countries, but in the latter two cases it goes further and incorporates the subsidiary into a vertically integrated supply and assembly chain. In these cases the trade is termed *intra*-industry trade (exchange of similar products) and, if it is conducted between subsidiaries of the same firm, *intra-firm* trade.

These distinctions are important because, while all these types of trade have a developmental effect for the regions in a trading system, *intra*-industry trade typifies the exchanges between the countries in the Triad and between developed regions, such as the countries of the EU, where economic regions are bound together in a single market. In contrast, a less-developed region would exhibit an inter-industry trade pattern – exporting products which embody its cheap labour, its low skill level or its raw material abundance. But there is quite a close association between FDI and *intra*-industry trade (Greenaway and Milner, 1987). A high level of intra-industry trade, especially of the intra-firm variety, signifies a close economic integration, probably of regions at roughly similar levels of development. The significance of this for the margins of the EU, whether the existing Mediterranean EU countries or the Accession countries in

Central Europe, will be drawn out fully below. In very broad terms, FDI in *inter*-industry trade probably would constitute a diversion of investment from Mediterranean to Central Europe, and a challenge to the trading position of the former. On the other hand, FDI in *intra*-industry trade might signify a tighter integration into European multinational production and a challenge not to the Mediterranean countries, but to the industrial core of northern Europe.

The Europe Agreements, asymmetric preferential trading arrangements established since 1990, set the framework for the creation of a free trade area between the European Community and each of the 'accession' states, those Central and Eastern Europe states wishing ultimately to join the EU (Flemming and Rollo, 1992). Insofar as they required accession states to adopt the economic constitution of the European Community, and stated the terms for undistorted competition and unhindered trade, they laid down something of a template for the transition of these centrally planned economies to a market system. Signing the agreements signalled a path for accession states' development, so significantly reducing the risk and uncertainty which foreign investors faced when contemplating investment in the CEE countries. The Europe Agreements have greatly facilitated the fundamental shift in CEE trade patterns: away from COMECON partners, towards Western Europe, which is further discussed below (Baldwin, 1994; EBRD, 1994, 1995).

In principle, foreign capital entering CEE economies should raise the level of competition and establish profit and efficiency objectives previously rare in state- or worker-owned enterprises. FDI can therefore help in the break-up of state monopolies, and, through the banking system, exert corporate control over industries which previously knew only the 'soft budget constraint' of state planning (Donges and Wiener, 1994). As regards trade links, opening up to world markets establishes world price relativities, helps introduce appropriate exchange rate arrangements, and assists a country to find comparative advantages in production which were previously obscured by wholly inefficient and economically costly central planning methods (Estrin *et al.*, 1997). It has been commonly accepted, then, that FDI and the opening of trade links to the West would be key instruments in fostering the transition of CEE countries to new, market-based structures (CEPR, 1990, 1992). The establishment of property relations, de-monopolization, privatization and the development of capital markets would be spurred on by the prospect of an inflow of foreign capital, which was much needed given the low levels of savings and acute shortage of investment capital in transition economies. Likewise, the injection of new technology, new marketing and management methods, a consumer orientation and access to Western supplies or Western markets could all be rapidly achieved in Central and Eastern Europe by FDI.

The Europe Agreements, and the prospect that at least Poland, Hungary and the Czech Republic from amongst the CEE countries may join the EU by 2004, change the dynamic of EU integration in many ways. Eastward enlargement necessitates fundamental reforms to the agricultural, budgetary and regional policies of the EU, as set out in the *Agenda 2000* proposals by the European Commission (1998). If for this reason alone, it has generated opposition by certain states such as Spain, which foresee higher budgetary contributions, reduced regional-fund revenues and a further, painful reform of the CAP regime. However, the Europe Agreements themselves, simply by instituting free trade,[1] have already changed the economic geography of the EU. They have shifted its economic centre eastwards, and admitted free exchanges in manufactures with a very low-wage region. Low-cost, labour-intensive goods imported into the EU from the East could compete directly with goods manufactured in the EU's 'southern' periphery and, of course, the region bordering Germany could well become integrated in an essentially German zone of production. One critical question is where Central Europe's comparative advantage lies: is it in low-skill, labour-intensive and materials-intensive production or is it in medium-skill manufactures (CEPR, 1992)? It is commonly said that the skill levels in Central Europe (principally Hungary and the Czech Republic – to a lesser extent in Poland) are high. If so, their present very low wage costs would attract medium and higher value-added manufacturing. They would not then constitute a direct economic threat to the 'south' but rather to countries of the heartland of the EU.

In short, the risk for the less-developed 'southern' EU countries is loss of export markets and a 'diversion' of FDI to the East, limiting their development potential within the EU. If there were 'diversion', this could denote a sharpening of inter-regional competition and a significant obstacle to the narrowing of regional disparities within the EU single market (Sheehy, 1994). The question of Central Europe's threat to the development of the southern EU, principally by eastward 'diversion' of FDI, is therefore highly pertinent (Baldwin, 1994) to European integration overall.

FDI and EU Integration as a whole

At an early stage a parallel was recognized between EU economic integration and FDI by multi-plant multinationals or transnational companies (Dunning and Robson, 1987). Both the organization of multinational firm and economic integration overcome barriers to economic activity and trade, and both build on comparative advantage and economies of scale for efficient resource allocation. Moreover, the creation of a single market from fragmented national markets can benefit the competitive multinational firm located within this widening market. Conversely, the creation of one market, in place of several fragmented markets, offers the opportunity for the competitive firm which enjoys economies of scale and a low-cost location to *concentrate* output at one location and export to the wider, unified market. In a single market, therefore, there is a tension between forces to concentrate activity and forces for its dispersal (Barrell and Pain, 1997). In very simplified terms these represent the opposing forces of economies of scale (and economies of agglomeration – reduction in firms' production costs from grouping together in a common production location) versus the transport costs of shipping from one location.

One might postulate the following reasons why a multi-plant operation should be adopted in a single market:

- plants will locate in locations that reduce costs (wage levels, other production costs, raw material availability) on the basis of comparative advantage and this will promote inter-industry trade;
- a firm with a specific asset, such as a brand or technology, may gain maximum market coverage by cloning manufacturing plants in several regions;
- because of regional differences between consumers a portfolio of plants may be needed;
- to achieve scale economies a manufacturer may concentrate production of different items of the product range in different locations ('horizontal' specialization) and exchange them as intra-firm, intra-industry trade;
- with scale economies a manufacturer may produce components in separate locations and exchange them for assembly ('vertical' specialization) as intra-firm, intra-industry trade;
- FDI may support exports by the creation of distribution and servicing channels;

- the liberalization of telecommunications and financial services, which was a key feature of the Single Market Programme, will generate FDI in service industries.

A range of studies has shown that, from the second half of the 1980s, the implementation of the Single Market Programme has been associated with a large jump in FDI activity within the EU (European Commission, 1988, 1996). Whilst FDI rose sharply throughout the OECD area in the period 1985–95, this was especially marked in the EU area (Barrell and Pain, 1997). This was an effect of the single market and of the concurrent capital market liberalization in the EU. At the beginning of the 1990s the EU absorbed 44.4 per cent of all world FDI, compared with 28.2 per cent in 1982–87. A comparison between FDI in the EU during 1984–85 and 1992–93 shows a five-fold rise by volume from all sources. In fact, over the period 1984–92, intra-EU FDI grew four times faster than did trade between EU countries (European Commission, 1996)!

In an early study (published later in 1994) Yannopoulos postulated, with supporting evidence, that the Single Market Programme's integration would generate a rise in FDI not only *between* member states, but also *from* non-EU FDI *into* the EU and from the EU *to the outside world*. The single market would attract non-EU FDI into the area for both 'defensive' and 'offensive' reasons. Offensive FDI is attracted by market growth and motivated by the desire to create a strong market share, while defensive FDI seeks to defend market share against encroachment from growing EU multinationals. Allied to this is the fear of exclusion from the single market by the operation of trade rules, such as anti-dumping suits ('contingent protection'). All this can help to explain the marked rise in Japanese and US investment in the EU since 1987. It can also explain a detectable 'diversion' of UK FDI from the USA to the EU and some diversion of investment away from Sweden and Austria in the very early 1990s (Pain and Lansbury, 1997).

Studies of German investment in the EU confirm Yannopoulos's picture. Germany is the world's fourth largest foreign direct investor, and the second investor in the EU after the USA. German FDI has risen very rapidly since the mid-1980s and in recent years three-fifths of these flows have been channelled to other EU states, implying that the EU share in the stock of total German FDI has doubled since the mid-1980s (Agrarwal, 1997). The increasing concentration of German FDI in the EU is therefore a striking feature. The econometric results contained in Pain and Lansbury (1997, p. 94) show that the rise in German FDI in the EU is strongly and significantly correlated with the implementation of the Single Market Programme. They estimate that the German stock of FDI in the EU may be $13.7 billion, or 17.5 per cent, higher than would have been the case without the single market. One should note, however, that the main beneficiaries of this large-scale German investment have been the UK, Netherlands and Italy. In contrast, flows to Greece were poor, and Portugal was not a prime beneficiary. Central Europe has come to take 10 per cent of German FDI flows in the 1990s.

Empirical data for intra-EU FDI since the mid-1980s also show that, in a sectoral breakdown, financial and business services predominated: 63 per cent of cumulative inflows were invested in services and only 31 per cent in manufacturing. To a considerable extent this can be explained by the liberalization of services in the single market agenda. In manufacturing, cross-border FDI reflected the comparative advantages of 'north' and 'south': investment in the northern states was in technology-intensive sectors (engineering, transport equipment, machinery); in the south it was in relatively basic products such as textiles, clothing, timber and wooden furniture (European Commission, 1996).

FDI in EU states of the 'periphery'

Relatively few studies have been devoted to the intra-EU distribution of FDI flows – whether originating from within EU states ('intra-EU' FDI) or from outside, particularly US and Japanese investment. In particular, data insufficiencies pose problems for a long time-series analysis – compounded by the fact that the EU itself has been progressively widened, in 1973, 1983, 1986 and 1995. A relatively early study by Molle and Morsink (1991) covered the period 1975–83. They found that the greater part of total inward and total outward FDI for EC countries (namely 60 per cent and 70 per cent, respectively) was with third countries (primarily to and from the USA) rather than *intra*-EU FDI. For intra-EU FDI the authors found that flows were concentrated very largely on the 'core' countries of the European Community. Italy, Spain, Portugal and Greece had only a small proportion of the total – though, at the time of the study, neither Greece nor the Iberian peninsula could have benefited as full EC members. The research showed that push factors (principally, the availability of funds and ownership advantages such as the high R&D expenditures of the investing companies) had a strong influence on investing country flows. The resistance to flows into recipient countries arose from cultural difference and especially distance (Molle and Morsink, 1991, p. 98).

More recent work by Dunning (1997) confirmed the picture of relatively small flows to southern European EU members, compared with flows to and between the core. But it did offer some important qualifications. Over the period 1980–95 the percentage share of total world inward FDI stocks fluctuated around 30 per cent for the 'core' of the EU (defined here as France, Germany, Italy, Belgium/Luxembourg, Netherlands and the UK), whereas the percentage going to the southern periphery rose from 3.9 per cent to 6.9 per cent. Within the EU itself the core generally held at least 80 per cent of the FDI stock. But the non-core countries' share did rise over time: the double influence of a large jump in the Spanish share and a large fall in the UK share, caused by the fall in the sterling–dollar exchange rate. Dunning shows that in the period 1985–93 the southern periphery (Greece, Spain, Portugal) received a very small share indeed of the total Japanese and US flows of FDI – with if anything a falling trend. On the other hand, the share of these three 'southern' countries as recipients of *intra*-EU total FDI flows doubled from 3.4 per cent to 8 per cent between the periods 1985/87 and 1991/93. But EU FDI flows (intra-EU and from outside) to the EU 'core' rose in turn from 26.4 per cent to 45.1 per cent of all EU FDI (Dunning, 1997, p. 12).

The picture of a small and declining Japanese interest in investments in the 'southern' periphery is confirmed in two works by Darby (1996a, 1996b). Before the early 1980s, Japanese manufacturing investments were relatively well established in southern Europe as a result of early FDI strategies. However, between 1983 and 1994, the share of the 'southern periphery' (Italy, Spain, Greece, Portugal) in Japanese total manufacturing investment in the EU fell from 39 per cent to 15.6 per cent (Darby, 1996a, p. 27). Their importance fell partly because of operational difficulties, but also as a result of the politicization of EC–Japan trading relations in the mid-1980s – particularly concerning imports of cars and IT equipment. In general, labour market inflexibility, the lower technological level and (paradoxically) the *less* protected nature of southern European markets for their products inhibited Japanese investments in the South, particularly in the new and more technology-intensive sectors. Japanese investment decisions were influenced by single market considerations; by the technological level of different areas; by a trade strategy to obtain access for Japan-made products and by Japanese firms' policies towards local transplant subsidiaries.

This picture of a very low, even if rising, share of the southern periphery in total world FDI in the EU, and within intra-EU FDI itself, points to a general conclusion. Investment flows have not helped the southern periphery to 'lift-off' to a rate of growth sufficient to close the output and income gap between them and the EU core. In making such a generalization, one should exclude the case of Spain, which in the first half of the 1990s has received a more significant share of intra-EU FDI. The question that poses itself, however, is whether eastward enlargement will 'divert' such FDI flows as there are towards the east, prejudicing even further the chances of the South's closing the development gap with the EU average.

It is not, however, self-evident that EU or other developed-country FDI flows into CEE will in fact represent a diversion from the other less-developed periphery of the Union. We must distinguish between investment in the tradeable and non-tradeable sectors of the economy. If we find investment in tradeable goods and services, is it motivated by the desire for local market penetration or is it export-oriented? Furthermore, will the goods invested in actually compete with those of southern Europe? At this point the inter-industry/intra-industry trade distinction becomes relevant. It is clear that, if the investment takes place in non-internationally tradeable sectors (such as infrastructure, real estate, hotel and tourism facilities or domestic services), FDI will not be competing directly with southern Europe. Capital will flow to CEE solely according to expectations of a high rate of return in a reconstructing, rapidly growing domestic market. What is more, even in the case of investment in tradeables, a local market orientation or longer-term 'strategic' market investments would not significantly affect existing trade patterns. Rather, in a multilateral world, by fostering growth, investment in tradeables would generate export opportunities for 'southern' EU countries.

Conversely, if FDI builds a CEE export capacity that is based on comparative advantage – promoting an outward processing trade or intra-industry trade with EU member countries generally – it may represent a diversion of potential FDI from southern European countries. The critical question here is whether the exports compete with those of southern Europe: whether, in other words, the comparative advantages of Central Europe overlap significantly with those of the South and therefore represent a direct trade challenge. If they do, Central Europe's low wage costs could attract FDI which previously might have flowed to the South.

Hence, one may postulate that investments in low-skill, raw-materials intensive branches of manufacturing will compete with the South, whereas investments attracted by the supplies of cheaper skilled labour, with experience in medium- or higher-value added manufacturing will tend to compete with the output and exports of the North. In a general way, therefore, *intra*-industry trade in mechanical and electrical engineering, in electronics and vehicles and vehicle components will compete with 'northern' output. By contrast, investments leading to *inter*-industry trade based on cheap, unskilled labour (for example in food processing, furniture-making, textiles, clothing and other light industry) may represent a diversion of FDI and may represent a challenge for southern output – which the South may or may not be able to meet.

CEE trade, FDI and closer integration with the EU

Quite early on, Central European countries managed to re-orient their trade away from the Soviet bloc and towards western Europe. Table 6.1 shows the structure of CEE5–EU trade over the period 1989–94. (The CEE5 comprise Czech Republic/Slovakia, Bulgaria, Romania, Hungary and Poland.) Over the period, total EU exports to the CEE5 grew at 24 per cent annually (Estrin *et al.*, 1997). It is clear that Germany is the dominant trade partner, accounting for 53 per cent of total EU exports to CEE5, with Italy in second place at 14.9 per cent. For its part, Austria has particularly close trading relations with Hungary and the Czech Republic. As will be seen later, there is a fairly close concordance between the FDI rankings and trade partner rankings, pointing to a correlation between trade and investment flows. Only in the case of the US, which may trade through its EU subsidiaries, is the FDI share far greater than its trade share. Overall, then, it appears that the Central European countries have experienced trade deficits with the EU because they have been importing machinery, materials and semi-finished inputs from the West. We must return below to the role of Germany, which is by far the biggest trader and the second largest investor in Central Europe. In recent years the share of Central and Eastern Europe in total German FDI and exports is approximately 10 per cent (DIW,

Table 6.1 EU trade with CEE, 1989–94 (US$ million)

	1989	*1990*	*1991*	*1992*	*1993*	*1994*
Exports						
Germany	6,884	11,464	11,904	15,093	16,092	19,137
Italy	1,459	2,162	2,378	3,490	4,283	5,379
France	1,161	1,348	2,011	2,582	2,355	2,761
UK	932	1,094	1,244	1,919	2,115	2,451
EU12 Total	12,627	18,618	21,658	27,802	30,231	35,934
Austria	1,625	2,309	2,891	3,442	3,527	4,114
Imports						
Germany	5,657	8,646	11,081	14,229	13,857	17,517
Italy	2,438	2,462	2,526	3,254	2,959	4,164
France	1,542	1,815	1,825	1,937	1,758	2,168
UK	1,223	1,268	1,140	1,333	1,528	2,051
EU12 total	12,401	17,398	20,159	24,524	23,717	30,753
Austria	1,536	1,885	2,207	2,608	2,509	3,125

Source: *IMF Trade Statistics Yearbook* (1995); Estrin *et al.* (1997).

1998; Deutsche Bundesbank, 1997a; *Handelsblatt*, 23 April 1998).

In a detailed analysis of the export reorientation and the growth of intra-industry trade, Hoekman and Djankov (1996) found that there has been little simple redirection to EU countries of Central European exports that had traditionally gone to COMECON: at most 20 per cent of export volumes were redirected in this way. Rather, export growth was either in products not previously exported to COMECON countries at all, or in traditional export items which had been significantly upgraded or differentiated. Bulgaria and Poland, the authors find, have experienced a greater change in the broad composition of their exports, while the Czech and Slovak Republics and Hungary have upgraded. In this latter group, intra-industry trade is more common, particularly as part of a vertical supply and assembly chain in which countries rely on EU firms for new machinery, components and know-how. Since CEE production is here integrated into supply and assembly, economic integration with Western European enterprises is clearly at its greatest in such cases. Hoekman and Djankov find that the strongest growth in exports is closely correlated with growth in such vertical intra-industry trade: particularly in the Czech and Slovak Republics, which also show the greatest reorientation in their trade patterns. Two questions emerge for the present discussion, therefore. How closely is EU FDI in the CEE countries linked to intra-industry trade patterns? And does the

Table 6.2 Direct investments in CEE, 1992–5, flows and stocks of FDI (US$ million)

	Flows				*Stocks*			
	1992	*1993*	*1994*	*1995*	*1992*	*1993*	*1994*	*1995*
Poland	467	1,109	1,095	3,659	470	1,614	2,840	7,843
Czech Republic*	1,003	568	862	2,558	1,606	2,053	3,077	5,797
Hungary	1,303	2,146	1,273	3,400	2,365	4,820	5,999	9,300
CEE10						9,448	15,222	21,483

*1992 data for Czechoslovakia.

Source: OECD (1996).

greater presence of intra-industry trade imply that the challenge of Central and Eastern Europe is less to southern than to 'northern' Europe?

Table 6.2 shows FDI flows and stocks in PCH (Poland, Hungary, Czech Republic) and in the 10 CEE states as a group. It is clear that, in recent years, following the turbulence and uncertainty of the first years of transition, there has been a rapid acceleration in investment inflows (and therefore of the stock of FDI). The CEE10, with a stock of $34.23 billion in 1996, represents 80 per cent of all FDI coming into transition countries. Within the group there is a high level of concentration, with PCH representing two-thirds of the total CEE10 investment stock. Inflows into Hungary were especially high in 1993 and 1995, as they were in Poland and the Czech Republic in 1995. Flows per capita have been highest in Hungary, which exerted a dominant role in attracting FDI in the early years. They are also high in the Czech Republic. FDI levels have been closely related to progress in transition (EBRD, 1994, 1995).

It is clear that there has been a rapid acceleration in investment inflows (and therefore of the stock of FDI) in more recent years, following the turbulence and uncertainty of the first years of transition. Estrin *et al.* (1997) rebut the claims by the World Bank (1996) that these flows to transition countries have been unimportant, pointing out that by both measures of FDI (inflow and per capita), PCH have achieved levels equal to, or greater than, the EU cohesion states. Only in the case of Poland is per-capita FDI relatively low for a country with such a large market and population. This must reflect the late development of its privatization programme and the earlier restrictive legislation on property ownership by foreigners (Kalinowski, 1996).

Table 6.3 Source of FDI stock in PCH, 1995 (US$ million)

	Hungary	*Poland*	*Czech Republic*	*Total*	*Percentage*
OECD total	8,833	7,593	5,575	21,991	96
US	1,490	1,883	787	4,160	18
Germany	2,297	1,518	1,739	5,554	24
Netherlands	981	1,338	787	3,106	14
Switzerland	271	378	821	1,470	6
France	758	432	542	1,732	8
Austria	1,484	384	316	2,184	10
UK	355	321	109	785	3
Italy	351	263	140	754	3
Total	9,301	7,843	5,797	22,941	100

Source: OECD (1997).

In terms of the investor countries, the US and Germany dominate, with Germany in prime position as both trader and investor. Table 6.3 lists the source countries for investment in PCH. Hungary is typical: the US and Germany account for more than one-half of FDI, while Austria lies in third position. In the Czech Republic, Germany accounts for 30 per cent of total FDI, and the US only 14 per cent. Their positions are reversed in Poland, where the US dominates, with 30 per cent of total FDI, and Germany holds only 10 per cent. Overall, EU FDI predominates in PCH, and within this, Germany plays the key role (DIW, 1996, 1997; DIW, Deutsche Bundesbank, *Zahlungsbilanzstatistik*). Turning to volume alone, in 1993–5 the stock of German FDI in CEE doubled, to DM 12.6 billion: including a doubling (to DM 4 billion) in the Czech Republic and a rise by DM 1 billion in both Poland and Hungary (Deutsche Bundesbank, 1997b). By 1996 German FDI flows to CEE countries as a whole reached DM 4.6 billion – 11 per cent of Germany's total FDI outflows, which was the same proportion as the CEE countries' share in Germany's goods exports (Deutsche Bundesbank, *Zahlungsbilanzstatistik*).

FDI and the form of integration in CEE countries

A range of studies, though with varied coverage and small samples, have looked at the motivations behind multinational investment in the CEE countries (EBRD, 1994). It is clear that there are a number of significant motives behind FDI in CEE (Lansbury *et al.*, 1996): the timing and scale of privatization programmes; the attraction of purchasing the previously very large market share of the state industry; market penetration and the strategic development of markets; cheap, skilled labour; trade links through the Europe Agreements; raw materials; tax incentives; a stable legal and macroeconomic framework. Lansbury stresses that privatization programmes, market penetration, a potential for intensive trade and a strong resource base are all powerful factors in econometric explanations of FDI flows, while relative labour costs remain as merely one important factor among several.

In a survey of 145 multinational companies in 16 countries from CEE and the former Soviet Union, Lankes and Venables (1997) make important distinctions according to the function of the FDI and the level of market transition which a country has reached. Countries at a more advanced stage of transition (such as the Czech Republic, Hungary and, to a lesser extent, Poland) are more attractive for export-oriented industries, while countries that are less advanced in transition attract investment seeking local market penetration or a natural-resource base. It is the macroeconomic and institutional stability, plus the cheap, skilled labour offered by the Czech Republic, Hungary and Poland that exert a strong pull. The export-oriented industries in Lankes and Venables' sample tended to be more 'upstream' in the production chain of the multinational (i.e. to create components and semi-finished goods). They sold 50 per cent of their output within the corporation itself, and obtained 30 per cent of their inputs there, which points to vertical integration in a supply chain and to some outward processing. Donges and Wiener (1994) suggest FDI of a similar character: they stress that, given the many qualified engineers and mechanics who have worked in the former military–industrial complexes of the CEE, cheap but skilled labour in the manufacture of components and investment goods is a prime attraction for FDI.

Details are given in Table 6.4 of the distribution of FDI in PCH by sector and branch. While the data series is incomplete, it seems that in the early years FDI focused very largely on manufacturing, while more recently services have become an important focus of interest.

The following points emerge:

- So far as manufacturing is concerned, food processing, consumer goods and vehicles production are very significant. The Czech Republic has experienced large investments in consumer goods, tobacco and automobiles; Poland has received much FDI in electrical and mechanical engineering; chemicals and food processing figure in all three countries.

Table 6.4 Sectoral and branch distribution of FDI stock in PCH, 1995 (US$ million)

	Hungary	*Poland*	*Czech Republic*	*Total*	*Per cent*
Agriculture and Fishing	110	19	5	134	0.6
Mining, quarry	81	21	81	183	0.8
Manufacturing	3,991	3,831	2,544	10,366	45
of which:					
Food products	1,124	1,007	825	2,956	13
Textiles, wood	457	793	88	1,338	6
Petrol, chemicals, rubber	1,070	759	275	2,104	9
Metal products, engineering	390	377	292	1,059	5
Office equipment Computers, TV	572	84	–	656	3
Vehicles, components	335	147	1,064	1,546	7
Electricity, gas, water	–	2	169	171	1
Construction	328	152	481	961	4
Trade and repairs	1,114	885	299	2,298	10
Transport, communications	836	172	1,415	2,423	11
of which: telecoms	709	63	–	772	3
Financial activities:	1,343	803	443	2,589	11
of which:					
Monetary institutions	–	527	354	881	4
Other financial institutions	–	241	..	241	1
Other financial, other insurance	–	266	89	355	2
Real estate, business activity	..	176	..	176	1
Other services	1,499	29	..	1,528	7
Unallocated	..	1,713	360	2,073	9
Total	9,301	7,843	5,797	22,941	100
Primary	191	40	86	317	1
Manufacturing	3,991	3,831	2,544	10,366	45
Services	5,119	2,259	2,807	10,185	44

Source: OECD (1997).

- Services investments have grown rapidly. Transport and telecommunications have received large investment in the Czech Republic; property and financial services are more important in Hungary and Poland; and construction has been the focus of interest throughout East–Central Europe.

Table 6.5 Branch distribution within CEE countries of German FDI, 1994 (DM million)

	Poland	*Hungary*	*Czech Republic*
All	1,105	2,792	2,770
Mining	x	–	–
Manufacturing	752	1,799	1,797
of which:			
Chemicals	110	114	124
Engineering	11	39	92
Vehicles	86	626	840
Electrotechnical	34	117	201
Distribution	272	443	437
Banks	x	240	228
Other lending institutions	–	x	10
Insurance	–	67	25

NB: x denotes confidentiality rule in data collection.

Source: Deutsche Bundesbank, Special Assessment (June 1996).

Table 6.5 breaks down the sectoral structure of German investments in Central Europe for just one year, 1994.[2] Two-thirds of investment goes to manufacturing, with large automobile investments in Hungary and the Czech Republic, a strong presence in chemicals and electronics throughout PCH, and a wider spread of investments in Poland. In fact, a particular characteristic of German FDI is the smaller size of German investing firms and their higher labour intensity compared with US investments. The large prestige investments by Volkswagen-Skoda (CR), Siemens and Daimler-Benz (H), Bahlsen and AEG in Poland, as well as heavy investments by the Commerz and Deutsche banks, are prominent exceptions to this.

This sectoral pattern raises few surprises. It reflects on the one hand infrastructural developments, strategic automobile investments, engineering, and, on the other, the pursuit of traditional comparative advantage in areas such as food processing and light industry. There is clearly a marked weighting towards medium valued-added manufacturing, requiring skilled employees in both the Czech Republic and Hungary – though in Poland some investments appear to have been undertaken in the more materials-intensive, low-skill branches. That said, data for this are significantly incomplete, reflecting underlying problems of data collection for balance-of-payments or central bank source data, host or home country statistics. There is also the 'unallocated' grouping in the statistics. Lastly, there is considerable uncertainty over the impact of joint ventures in the data. They may be very important in outward processing and other production agreements; but because the capital injection is small, they may be under-represented in the statistics.

Building on a range of case studies, Estrin *et al.* (1997) concluded that there are marked differences between the pattern and motivation of US investments and those from Germany (or Austria). US FDI tends to be concentrated in a few large projects which seek market penetration and benefit from scale economies. They reflect American acceptance of a higher risk/return ratio and the superiority of US technology, managerial and marketing expertise. US enterprises also have invested heavily in utilities and infrastructures. By contrast, German investments tend to be conducted by smaller firms, with a higher labour intensity than US firms. A multiplicity of small and medium-sized enterprises (SMEs) from Germany have forged cross-border joint ventures – for outward processing, for component manufacture or as a first step to market entry. There are many cases of Bavarian SMEs in the Czech Republic, and of medium-sized firms from throughout Germany forging joint ventures in Poland. For the reasons given above, such investments are poorly captured by the statistics.

Given their long history of trading with eastern neighbours (and, since the late 1970s, outward processing there), German firms have considerable informational advantages and they have cultural affinities in Central Europe. With little geographical distance and low cultural barriers, Germany is a natural investment partner (Agrarwal, 1997). Such firms can exert strong managerial and logistical control, and of course they have considerable support from their government and banking systems. Germany is well placed to build on its distribution and service links. With continuing scope for privatization in utilities, railways, banks and the so-called 'strategic industries', there remains a high potential for German investment, attracted at least by the expected income growth in these markets (Agrarwal, 1997).

There is some dispute as to whether German firms are motivated by cost considerations to escape the high production costs in Germany. The Deutsche Bundesbank (1997b) stresses the search for cost reduction investments in labour-intensive components and materials, as a strategy to reduce German production costs. This would imply a redistribution of production locations on the basis of comparative advantage. The host country would gain, among other things, access to Western supplies and markets. This mode of entry allows German companies facing high costs at home to maintain close organizational control while benefiting from wage levels one-tenth of those in Germany. But, to the extent that investment belongs to an outward processing strategy, the production site is fitted into a supply chain offering low-cost labour without developing the host country's sourcing of materials. In consequence, the spillover gains to the host country are obviously small.

It is difficult to assess how common such outward processing agreements are among German SMEs because the data are limited and give weak coverage to joint ventures. Other studies of German FDI tend to rank the search for cheaper wage costs lower in the motives of German investment in CEE. Barrell and Pain (1997) distinguish the motivation for German FDI *within* the EU (which is more sensitive to cost factors) from that for FDI *outside* the EU, which is attracted by the existence of a larger market and demand for specialized German products. Beyfuss (1996) showed that penetrating or preserving markets played a dominant role in German FDI, with wage and other cost motives ranking lower. Agrarwal (1997) restates this view, and insists that German FDI in Central Europe is market- rather than cost-oriented. He goes further: to explain that there is little evidence of German FDI diversion from southern Europe to Central Europe, since market-seeking FDI tends to create, rather than divert investment from other target countries. Most German FDI seems to have been financed not by diversion from other host countries, but by the mobilization of additional equity capital (Agrarwal, 1997, p. 108).

Is FDI 'diversion' reshaping the southern to eastern margin of Europe?

From the evidence given above, it may be that there has occurred a small, possibly an insignificant amount of FDI 'diversion', primarily to Poland. Clearly, in the past ten years, there has been a rise in the small share that the 'southern' periphery receives in intra-EU FDI. But this has been dwarfed by the growth in intra-EU FDI among the core countries, in response to the single market programme, which freed capital movements. In the 1990s, on the other hand, there has been a very large growth in foreign investment in CEE, with a doubling of German FDI in Poland, Hungary and the Czech Republic in the period 1993–95 alone. Germany and the USA are the key investors, followed by Austria. However, to keep a sense of proportion, it should be said that the volumes and *per-capita* levels of FDI in Central Europe still only equal those in the cohesion states of the EU. Even this applies less in Poland, where FDI is smaller than the size of the country and its population might warrant.

To establish the impact of these investments, however, and whether they compete with the 'South', we must assess the branch of activity in which they have been made, and consider the motivation for them. The large share of investment in local services and other non-tradeables is readily apparent from the data: telecommunications, transport, property and financial investments occupy a major place – though the exact nature of the banking and finance investments would require further specification. Within the tradeable-goods sector, the branches favoured by foreign investors have been vehicles, chemicals, electronics, mechanical and electrical engineering, consumer goods, food processing and tobacco. Some interpretation is needed to decide whether these investments are export-oriented or conducted with a view to local-market penetration and strategic market development. The weight of the investments in tradeable manufactures appears to be directed to intra-industry trade. The preponderance of such investments in engineering, chemicals and electronics points to component and assembly activities, as well as to final product manufacture. Whereas investments in food processing, tobacco and some consumer goods could compete directly with low-wage, low-skill exports from southern Europe, the pattern that emerges for FDI going to CEE for the manufacture of tradeables suggests competition with 'northern' rather than 'southern' products.

The motivation for investment, certainly as far as German manufacturers is concerned, suggests that market development and penetration are the objectives, rather than pure cost reduction. However, even in cases where the joint ventures and investment by German medium-sized firms may indeed be motivated by relative labour costs, the firms are seeking skilled labour. For a number of reasons, that would not in any case have motivated them to invest in southern Europe. The studies of German investment described above all show that the expansion into Central Europe is a natural move for German firms; whereas German manufacturing investments in Greece and, to some extent Portugal, have always been sparse.

The fact that German firms have raised equity to fund their CEE developments, rather than transferring existing funds for investment, is one further sign that there has been little 'diversion' of capital from southern Europe. As to competition from the new output generated by FDI, one has also to bear in mind that the greater part of the CEE foreign investment is not in any case in tradeable manufactures, and so cannot be said to be a competitive threat. Finally, it would seem that (with the partial exception of Poland) the foreign investments in manufacturing tend to compete with the trade of 'northern', rather than 'southern' countries.

Notes

1 Subject to special treatment for 'sensitive products' and contingent protection in the form of anti-dumping rules (see Fleming and Rollo, 1992).
2 Such data is rarely published by the German Central Bank.

References

Agrarwal, J. P. (1997) 'European Integration and German FDI: Implications for Domestic Investment and Central European Economies', *NIESR Economic Review*, **2**, 110–11.

Baldwin, R. (1994) *Towards an Integrated Europe* (London: CEPR).

Barrell, R. and Pain, N. (1997) 'The Growth of FDI in Europe', *NIESR Economic Review*, **2**, 63–75.

Beyfuss, J. (1996) 'Erfahrung deutscher Auslandsinvestoren in Reformländern Mittel- und Osteuropas', *Beiträge zur Wirtschafts- und Sozialpolitik*, 232. Institut der deutschen Wirtschaft (Köln: Deutscher Institutsverlag).

Caves, R. E. (1996) *Multinational Enterprises and Economic Analysis*, 2nd edn (Cambridge: Cambridge University Press).

CEPR (1990) *Monitoring European Integration: The Impact of Eastern Europe* (London: CEPR).

CEPR (1992) *Monitoring European Integration: Is Bigger Better? The Economics of Enlargement* (London: CEPR).

Darby, J. (1996a) 'Less Successful Strategies: Japanese Investment in Southern Europe', *South European Society and Politics*, **1**(1), 24–46.

Darby, J. (ed.) (1996b) *Japan and the European Periphery* (London: Macmillan).

Deutsche Bundesbank, *Zahlungsbilanzstatistik*.

Deutsche Bundesbank (1996) Sonderauswertung der deutschen Direktinvestitionen, *Beilage zur Reihe der Zahlungsbilanzstatistik*, **6**.

Deutsche Bundesbank (1997a) 'Deutsche Zahlungsbilanz im Jahre 1996', *Monatsbericht*, **3**, 47–60.

Deutsche Bundesbank (1997b) 'Die Entwicklung der Kapitalverflechtung, 1993–1995', *Monatsbericht*, **5**, 63–77.

DIW (Deutsches Institut für Wirtschaftsforschung) (1996) 'Ostmitteleuropa auf dem Weg in der EU-Transformation, Verflechtung, Reformbedarf', in

Beiträge zur Strukturforschung, Heft 167, Kapitel 2 (Berlin: Duncker & Humblot).
DIW (1997) 'Ausländische Direktinvestitionen in den Transformationsländern', *Wochenbericht,* **11**, 183–9.
DIW (1998) 'Starke Ausweitung des Aussenhandels mit den Reformländern MOEs von 1992 bis 1997', *Wochenbericht,* **7**, 142–8.
Donges, J. B. and Wiener, J. (1994) 'Foreign Investment in Eastern Europe's Transformation', in V. N. Balasubramanyam and D. Sapsford (eds), *The Economics of International Investment* (Aldershot: Edward Elgar), pp. 129–48.
Dunning, J. (1993) *The Globalization of Business* (London: Routledge).
Dunning, J. (1997) 'The European Internal Market Programme and Inward Investment', *Journal of Common Market Studies,* **35**(1), 1–30.
Dunning, J. and Robson, P. (1987) 'Multinational Corporate Integration and Regional Economic Integration', *Journal of Common Market Studies,* **XXVI**(2), 103–25.
EBRD (1994, 1995) *Transition Reports* (London).
Estrin, S., Hughes, K. and Todd, S. (1997) *Foreign Direct Investment in Central and Eastern Europe: Multinationals in Transition* (London: Pinter).
European Commission (1988) 'The Economics of 1992', *European Economy,* **35** (Brussels).
European Commission (1996) 'Economic Evaluation of the Internal Market', *European Economy, Reports and Studies,* no. 4.
European Commission (1998) *Agenda 2000,* http://europe.eu.int, DGVI-Agenda 2000-EN.
Flemming, J. and Rollo, J. M. C. (1992) *Trade, Payments and Adjustment* (London: RIIA/EBRD).
Greenaway, D. and Milner, C. (1987) 'Intra-industry Trade: Current Perspectives and Unresolved Issues', *Weltwirtschaftliches Archiv,* **123** 39–48.
Handelsblatt (various).
Hoekman, B. and Djankov, S. (1996) 'Intra-industry Trade, Foreign Direct Investment and the Reorientation of East European Exports'. CEPR Discussion Paper no. 1377, April.
Kalinowski, J. (1996) 'Ausländische Direktinvestitionen in Polen-Anfang eines dauerhaften Wachstums?', *DIW Vierteljahresheft zur Wirtschaftsforschung,* **2**, 248–55.
Lankes, H-P. and Venables, A. J. (1997) 'FDI in Eastern Europe and the Former Soviet Union: Results from a Survey', in S. Zecchini (ed.), *Lessons from the Economic Transition* (Dordrecht: Kluwer), pp. 555–65.
Lansbury, M., Pain, N. and Smidkova, K. (1996) 'FDI in Central Europe Since 1990: an Econometric Study', *NIESR Economic Review,* **5**, 104–14.
Ministère de l'Économie et des Finances (1996) 'L'investissement étranger en Europe centrale et orientale, vecteur de transition, enjeu pour les entreprises françaises'. Notes bleues de Bercy, 16 October.
Molle, W. and Morsink, R. (1991) 'Intra-European Direct Investment', in B. Bürgenmeier, and J. L. Mucchielli (eds), *Multinationals and Europe 1992* (London: Routledge), pp. 81–101.
OECD (1997) *International Direct Investment Statistics Yearbook* (Paris: OECD).
Pain, N. and Lansbury, M. (1997) 'Regional Economic Integration and FDI: German Investment in Europe', *NIESR Economic Review,* **2**, 87–99.

Sheehy, J (1994) 'Foreign Direct Investment in the CEECs', in *The Economic Interpenetration between the EU and Eastern Europe*. European Economy Reports and Studies, 6.

World Bank (1996) *World Development Report on Transitional Economies* (Washington, DC: World Bank).

Yannopoulos, G. N. (1994) 'Multinational Corporations and the Single European Market', in V. N. Balasubramanyam and D. Sapsford (eds), *The Economics of International Investment* (Aldershot: Edward Elgar), pp. 329–45.

7
Transnational Planning in the German–Polish Border Region

Ann Kennard

The significance of borders

The map of Europe has changed greatly since the revolutions of 1989–90. In some regions, for example in former Yugoslavia, the actual borders have moved. But in others, it is the significance of the border, the relationship between the countries on either side, which has changed. This is the case between Germany and Poland. On the one hand, the unification of the two German states moved the external border of Germany – and of the EU – towards the east. On the other hand, the more general collapse of the Eastern Bloc undermined the bigger, east–west division of Europe, which Poland's western border now marked. With this marginal situation in mind, it is the intention here to investigate the changes in relations between border actors in the two countries, to evaluate the level of openness of this border, and to consider whether attempts to integrate planning mechanisms in the transborder region are likely to be successful in the future.

The concept of the border has long been the subject of much analysis and debate. Seemingly mere lines on a map, as a geographical entity or symbol, borders are in reality much more than this. They are the most recent tangible manifestation of historical development, political or military conflict, demographic movements, and cultural, social and economic division. Borders can have many different characters. Medieval Europe, for instance, developed hereditary fiefs, where the lord granted land to his vassals as a reward for loyalty which was then passed on from generation to generation. The concept of natural, physical boundaries, on the other hand, which has existed for thousands of years, was used as recently as the end of the Second World War, when the Oder–Neisse line was set up, albeit by force.

Lord Curzon, the late-nineteenth-century diplomat, who as Viceroy of India was called upon to organize and conduct the proceedings of five Boundary Commissions, made known his ideas on boundaries in his famous lecture of 1907 (Curzon, 1908). He perceived no great difference between frontiers and boundaries, but he recognized the process by which a frontier may become a demarcated line. He knew the need for force along frontiers, but felt that boundary-making was the domain of 'strong, wise, just imperial powers' (Jones, 1959, p. 250).

Ladis K. D. Kristof attempted to clarify our theoretical understanding of borders with a very thorough investigation of the etymology and the historical and political attributes of borders as 'frontiers' and 'boundaries' (Kristof, 1959). He points out that other languages, such as French and German, do not have this distinction ('frontière' and 'Grenze'). In the English language, *frontiers* are seen as outer-oriented, integrative: a 'zone of transition from the sphere (ecumene) of one way of life to another' (p. 273), neither of which sees itself as self-sufficient or autonomous. *Boundaries*, on the other hand, are inner-oriented, a separating factor between two self-sufficient forces 'created and maintained by the will of central government' (p. 273).

Yet there is general agreement amongst those writing and researching in the field that, due to differing sociopolitical structures, cultures and ideologies, boundaries inevitably create zones of friction. Loyalties and duties to one's 'own' society and government mean that disagreements, and occasionally worse, will flare up from time to time. On the other hand, 'borders' may be seen to include both the legal borderline between states and the frontier of political and cultural contestation which stretches away from the borderline (O'Dowd and Wilson, 1996, p. 2). In this interpretation the frontier transcends the borderline, and encompasses the economic, social and political landscape of the borderlands' people.

The situation in Europe since World War II has been a mixed and confused one where borders are concerned. From the geopolitical chaos unleashed by Hitler's megalomaniac ambitions emerged a Europe divided down the centre between a new Soviet-influenced Communist East and a capitalist West oriented towards the United States. Within these two blocs many borders had been shifted for differing political reasons. In the post-war environment one would have to regard those as jealously guarded *boundaries*, rather than outward-looking *frontiers*, even within their respective blocs.

The gradual emergence of integrated structures in the western half of the continent superimposed porosity between the member countries of the European Communities. But the peoples there feel nevertheless that they owe their overwhelming loyalty to their nation-states. The situation in the East was yet more pointed: so-called 'integration' had been forced on those countries via the Soviet-led Communist 'blanket'. All the more reason for those subjugated nations to retreat into a national cocoon. Here, however, they did not do that in loyalty to their (Communist) national government, but to some other institution like the Church or something rather more intangible, such as traditional cultural, family and social values.

If borders in the East remained largely closed and the populations inward-looking, the situation in the West gradually changed, with structural integration on the one hand, and US-led economic and cultural globalization on the other. Still the concept of the *frontier*, with its openness, is perhaps more of an American idea. It may not really be appropriate to the European context, since it is generally considered to be marked by some element of rebelliousness (from the national centre) and an absence of laws. One of the accepted characteristics of 'frontierism' may nevertheless be relevant; namely, that the inhabitants of borderlands often develop distinctive interests which may have little in common with those of their central government.

As the likelihood of conflict in Europe as a whole diminished it was perhaps inevitable that, in the West at least, there would be a certain 'blurring' of the post-1945 borders. There had been massive movements of population across all borders, which resulted in significant overlaps between ethnic and/or religious minorities. Thus the idea of a *boundary* as an exclusionary device became inappropriate in a Europe where nations and states cannot always be said to coincide (O'Dowd and Wilson, 1996, p. 4). In addition, the general trend towards globalization engendered the feeling that trans-border cooperation might well lead to synergies which would benefit both sides. Indeed, since regions close to their national borders are often disadvantaged in economic terms due to their distance from centres of national economic activity, their very marginality suggests a reorientation towards the other part of their region, across the border.

This perception, together with the acceptance of a certain 'supranationalism' in terms of political and economic interests (NATO and the European Community), led in Western Europe to the establishment of bilateral and multilateral commissions whose aim has been to facilitate the coordination of national policies that have international ramifications. These international bodies have developed most rapidly in the core area of the European Union and include the French–German–Swiss, French–German, Dutch–German and Belgian–German Planning Commissions. In all such commissions, representatives of central, regional and local governments meet to discuss ways of coordinating planning decisions taken within common border regions (Scott, 1996, pp. 86–8). The Council of Europe has also played a role in the development of a European border region policy, aiming to harmonize spatial planning and regional development policy, and more generally supporting the principle of transboundary regionalism. The principle of transboundary cooperation was eventually enshrined in a Council of Europe agreement in 1980.[1]

Lowering frontiers in the West – along with the growth of the Single European Market, plans for future political union, and, more recently, the establishment of a security barrier around the edge of the EU (Schengen) – have tended to create a higher barrier against the outside world ('Fortress Europe'). This was happening even as Gorbachev was returning the powers of self-determination to the countries of Central Europe. It is ironic that as the boundaries between nation states in the West became blurred, those in the East were emerging into sharp relief. What might come out of this mix of differing experiences and different trajectories at the East–West boundary itself?

The German–Polish border: a paradigm margin at the centre of Europe

Of all European border regions, the German–Polish border may be said to be the one which demonstrates the greatest multiplicity of marginal attributes. Politically a number of conflicting influences may be discerned, all of them imposed externally, and all contributing at once to a sense of insecurity and defensiveness in the regions on either side of the border. The location at the front line between two implacably opposed political and defence systems, and the need on both sides to find a national identity in a chaotic post-war situation made this, in today's terms, probably the 'hardest' border in Europe.

From the outset in 1945, the 'Oder–Neisse line' was implicitly part of a new rupture between East and West. When the East claimed part of Germany too, the line became a merely 'internal' border of Comecon and the Warsaw Pact – though the inevitable enmity between the Poles and Germans (in this case, the GDR) ensured that it still retained its exclusionary role. Another element of marginality was literally imported into this border region: 1.5 million ethnic Poles expelled from their own eastern territories when the Polish borders moved westwards under agreements struck at Yalta and Potsdam. These people, deprived of their homes in areas which Stalin had claimed for the Soviet Union, came to live in houses vacated by Germans expelled or fleeing from Silesia and Pomerania. For, at the behest of Stalin (and with the agreement of the Western Allies), these areas were now in the West under 'Polish administration'.

Thus there were a considerable number of dislocated groups scattered about the new western regions of Poland, living alongside a minority of Germans who had resisted leaving their homeland for the new Germany to the west. Throughout the Cold War, the juxtaposition of these ethnic groups was largely ignored for political reasons. Then Brandt's *Ostpolitik* attempted to find a *modus vivendi* between the two Germanies and reconciliation with East Europe by accepting the border arrangements agreed at Yalta. This brought about a certain thaw in the 1970s (Kennard, 1996, p. 110). Though the western border with Poland was not, of course, contiguous to the Federal Republic at the time, there were continuing West German territorial claims to post-war Polish territory. Once in office, Willy Brandt sought Poland's forgiveness for this stance and, in 1970, signed the German–Polish Treaty, which accepted the Polish western border as a permanent and inviolable part of the map of Europe.

Although it was in one sense an 'internal' border to Comecon and the Warsaw Pact, seen from the respective national perspectives, Poland's western border was an ethnic and cultural dividing line: very much the outer edge of the 'self' faced by the 'other' on the opposite side. In spite of official political and economic links, in reality the divide between the GDR and Poland was close to total. Culturally also there was never much evidence of common themes or activities. The advent of *Solidarnosc* in Poland in 1980 further underlined the difference between the two societies. The

GDR closed the border to keep out this Polish 'bacillus'. The economies drew apart as the strikes called by *Solidarnosc* damaged the East German economy (and others): essential supplies of Upper Silesian coal all but dried up during this period. Thus the abrupt change in 1989–90 had a profound effect on the consciousness and activities of the inhabitants on either side of the border.

It is all the more surprising, then, that today this border is sometimes termed a 'border of affluence' (Macków, 1995, p. 34). This is especially unexpected given the disparity between the relative wealth of the ex-GDR Germans, subsidized by the West Germans, and the Poles, who are in a largely agricultural area and only just emerging from serial economic difficulties caused by the inefficient centrally planned economy followed by the unemployment and other hardships of the economic 'shock therapy' of the first post-1990 government. Nevertheless, economic activity in the regions surrounding the border is significant and growing very fast.

There has been a rapid response of local and unofficial actors to the post-1989 situation, ahead of any official government-backed institutions. This has led to a huge expansion of private trans-border trade, sometimes called the 'bazaar economy' (Stryjakiewicz and Kaczmarek, 1997, p. 4; Krätke, 1997, p. 3), the biggest Polish bazaar, in Slubice opposite Frankfurt (Oder), containing 1,000 stalls.

The Polish Ministry of the Economy has quoted the turnover of the trans-border economy at US$7.1 billion, equivalent to no less than 29 per cent of Polish exports (Slojewska, 1997, quoted in Stryjakiewicz and Kaczmarek, 1997, p. 4). This trade points up the difference in incomes in the two countries; for it relies on huge numbers of relatively wealthy German customers hoping to profit from the low-priced goods. The difference in incomes between Poland and Germany as a whole is estimated at approximately 1:10 (Stryjakiewicz and Kaczmarek, 1997, p. 2).

But the situation is not static: the dramatic loss of jobs and massive de-industrialization on the German side of the border compare rather negatively with the relatively stable employment position on the Polish side (Krätke, 1997, pp. 5–6). Extremely different futures are envisaged for the economic development of the whole border region: as a 'drainage area' on the outskirts of the EU; as a stagnating hinterland of the metropolis of Berlin; or, at the other extreme, as a future 'tiger' region with a high percentage of low-wage industries (Krätke, 1996, p. 648).

The idea of a low-wage economy alongside a wealthy partner as 'provider' suggests a development scenario similar to the *maquiladora*[2] on the US–Mexican border, where both US and Mexican border populations are highly urbanized, but the Gross Regional Product (GRP) is massively differentiated. In 1994 the County of San Diego had a GRP of $67.1 billion, whereas that of the Municipality of Tijuana was around $3 billion (Ganster, 1996, p. 178). Migration is an important factor in such a case: swift demographic expansion has provoked infrastructure crises in border cities, particularly on the Mexican side where resources are inadequate. Yet the presence of Hispanic populations on both sides of the border as a result of the migration has encouraged strong trans-border social and cultural links.

Bearing in mind the numbers of Poles who are already commuting across the border (albeit mainly to Berlin), it is possible that a similar scenario could develop here. However, anything which assumed the permanent subordination of the Polish to the German economy would be unlikely to be acceptable as a way forward for Polish border-region development. Thanks to their historical relationship, and the fact that the Polish economy is currently one of the fastest growing in Europe, the interface between the Polish and German regional economies will, in any case, be the subject of a continuing discussion. If the right approach is taken to the future development of the region as a whole, there is no doubt that synergies between the regional economies are very possible.

After central planning was consigned to history in central Europe, regional identity started to re-emerge and regions needed to be redefined. Whilst a debate commenced on the reorganization, that is, decentralization of the regional administrative structure in Poland, the initiative to cooperate across their common border came from Germany. There were many problems transcending the border which were badly in need of a mutual approach. These included a lack of regional and local development potential, the small number of border crossings restricting the flow of trans-border traffic and divided twin cities which were without common utilities and infrastructure. The Oder river was also being neglected as a possible waterway and neither side felt responsible for the ever-increasing environmental problems (Gruchmann and Walk, 1996, p. 129). The result, after a surprisingly short period of transnational negotiations, was the establishment between 1991 and 1995 of four Euroregions covering the full length of the border from the Baltic to the Czech Republic: *Pomerania, Pro Europa Viadrina, Spree–Neisse–Bober (Sprewa–Nysa–Bóbr)* and *Neisse (Nysa/Nisa)* (see Figure 7.1). *Neisse* was the first established of these Euroregions and includes also the Czech border region.[3] The last, *Pomerania*, stretches as far as the Baltic coast in the north and, since February 1998, has also included the most southerly region of Sweden, Scania (see Figure 7.1).

The institutional framework of cooperation

It is useful at this juncture to examine the institutional framework of cooperation across the German–Polish border and to set it in the wider European context of regional policy and planning. Since the stimulus for cooperation in the case of the German–Polish border came from the German side, and the source of potential funding was the EU, the institutional framework was therefore transferred also. This was an ambitious and rather risky enterprise, since it effectively imposed structures and bodies from the EU and Germany upon a nascent Polish parliamentary democracy, which had no real chance to question their appropriateness. In the event, the arrangements seem to have worked reasonably well so far.

The EU's 1994 document *Europe 2000+*, which divided the Union for regional planning purposes into a number of transnational 'study regions', made clear that the European Commission's perception of future regional policy and planning was decentralized and transnational. The development of special funding mechanisms, as well as institutional frameworks, aimed at encouraging cooperation across the borders of EU member states demonstrated the intention to move regional policy decision-making down to a subnational level in order to solve problems and conflicts not always perceived, or appreciated, by central governments.

Now that this cooperation has been extended across the external border with East–Central Europe – itself a clear break with the past for Poland and Germany – there is an extra dimension, which goes beyond the attempt to influence regional development policy. Cooperation in this part of Europe represents a clear statement of intent on both sides concerning the future development of the bilateral relationship between the EU member state and the East–Central European neighbour. Poland, Hungary and the Czech Republic are set to become an integral part of the EU, and this cooperation will help in the process. Responsibility for spatial development and planning in the border regions of Europe lies with a multitude of different bodies, and these are gradually being adapted to cover the external borders in anticipation of future Central European membership. Four different levels of policy-making can usefully be distinguished (Roch *et al.*, 1998, p. 7).

At the top level, *Europe-wide* regional development concepts emerge from policy decisions from the Council of Ministers, the Commission or the European Parliament in the form of documents such as *Europe 2000+*. There is also an informal grouping of ministers responsible for spatial planning, which can make recommendations. Other help and recommendations come from the Council of Europe, and from the unofficial but very influential Association of European Border Regions (AEBR) based in Gronau, where the first Euroregion, 'Euregio' (between Germany and the Netherlands), was born.

At *national* level, inter-governmental commissions play the official leading role and generally set the agenda for cooperation, but only in an advisory capacity. For specific purposes they set up committees and commissions. Currently the most important for the German–Polish border region is the Spatial Planning Commission – a body to make concrete recommendations on trans-border regional development planning, which was transferred intact from Euregio. The Commission has decided to make recommendations in three areas of cross-border cooperation: setting up local-authority development plans; carrying out regional and inter-regional planning measures, including the exchange of information; and exchanging of information or consulting about investments that are significant for spatial development in the border region (Roch *et al.*, 1998, p. 9).

In Germany the *Länder* have a role to play at a subnational or broader *regional* level, but they work in concert with a variety of specialist associations. It is the *Länder* which are responsible for the regional development plan *(Landesentwicklungsplan)* and they are also involved in putting up (or procuring) the matching finance, which is a condition of EU funding of cross-border projects (see below). Regional planning in Poland is decidedly in its infancy. Polish political life has always been very centralized, both before and during the Communist period, so that there is much work to do on the synchronization of any trans-border planning. So far, some of the regional input, from the *województwa*,[4] has had a rather negative impact, since the '*wojewoda*' is a personal nominee of the Prime Minister and, as such, really a channel for central government policy. The main problem for any trans-boundary cooperation is that, in the national context, the western border region of Poland

Figure 7.1 Euroregions on the German–Polish border

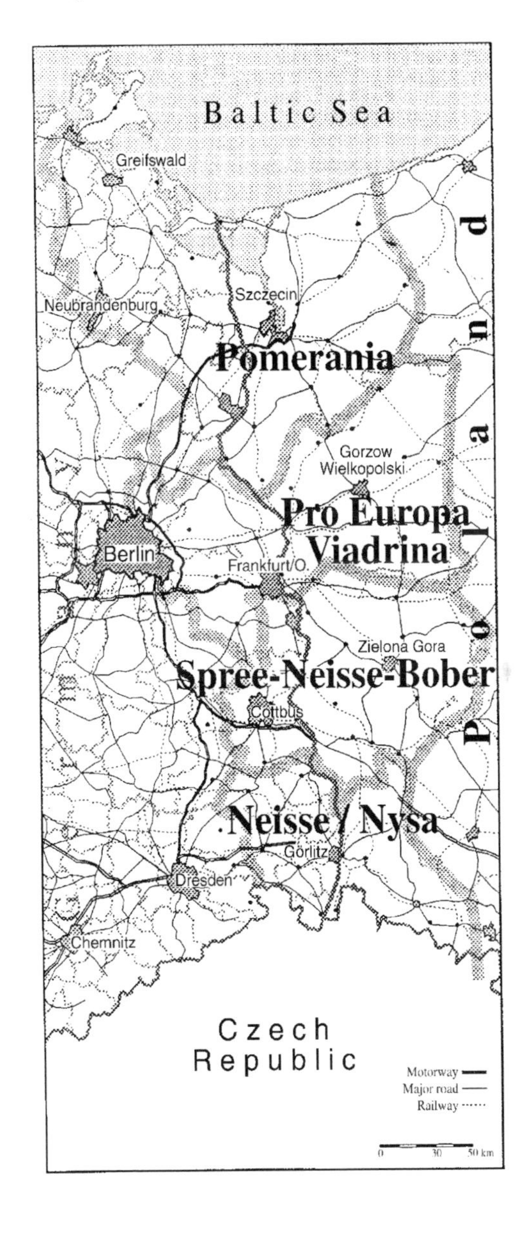

is perceived as relatively affluent compared with many other regions of Poland – including, of course, the border to the east. Thus little that is helpful or constructive has come down from Warsaw via the *wojewoda*, due to a possible fear of collusion with the West in this national margin, as it pursues the region's economic potential.

However, throughout this decade it is on a *local* level that activities have been at times their most intense and fruitful. Euroregions are by now probably the most well-known cross-border organizations dedicated specifically to trans-national planning and cooperation. They are generally composed of various local groupings and associations, both public and private, from the countries either side of the border in question. One of the prime aims is to reduce locational disadvantages, especially in relation to unemployment, by creating jobs and halting emigration from the region (Gruchmann and Walk, 1996, p. 132). Due to their historical and geopolitical position in Europe, the German–Polish Euroregions have a resonance not shared by any others east or west. The fact that they were established so soon after 1989 is a tribute to the pragmatic approach of the actors involved. That there were misunderstandings and hold-ups along the way is hardly surprising in the circumstances. What is surprising is that, against such a bleak background of historical memory, so much has been achieved in such a short time.

Euroregions as vehicles for trans-border problem-solving

The *raison d'être* and function of Euroregions is broadly to encourage trans-border cooperation in such a way as to minimize disadvantages inherent in peripherality, whether in the labour market, in the overall economic infrastructure, transport and communications, or, last but by no means least, in environmental matters. There are also positive advantages to be gained from trans-border cooperation in fields such as education, sport, cultural events and amenities. The question now is whether the Euroregions are the appropriate vehicle for the successful achievement of such cooperation.

Euroregions are international partnerships, dedicated to solving otherwise intractable trans-boundary problems, like those above. The central coordinating bodies include representation both from public authorities of the countries concerned – municipalities, towns and districts – and from regional associations of an unofficial, that is non-statutory, nature. Therein lies one of their main problems: that Europeans do not have a mandate for the implementation of any cross-border cooperative activity. There are now more than 60 Euroregions which have grown up and developed over a period of almost 40 years across the borders of Europe. Although they are formally instituted and, without exception, managed by local, regional or even occasionally national statutory authorities, they are in fact voluntary bodies which can act only in an advisory capacity. Nowhere in Europe has it yet been legally possible to set up any trans-border institutions under public law (Gabbe, 1992, p. 92). Each Euroregion develops differently, since they all have a different *raison d'être*.

Take, for instance, the original *Regio Basiliensis*, established in 1963 as a planning institute for the city of Basel and subsequently complemented by an international coordinating service, with groups in South Baden (Germany) and Haut-Rhin (France) linked into it. Hans Briner, former Director General of the *Regio*, has openly admitted that the organization was a necessary instrument to ensure that Basel, a large city in Switzerland, was not rendered economically and politically isolated on the periphery of the EU (Scott, 1996, p. 85). Though it has now been superseded by an inter-governmental commission, this grouping still exists in order to keep the formal institutions of cooperation on their toes (Anderson, 1982, p. 8).

More recently, the Euroregion *Neisse* was formed on the eastern border of the EU very soon after the 1989 revolutions, at the very beginning of the East–West rapprochement. It had a quite different purpose: to try to solve the environmental problems caused in particular by the chemical industry in that region, the so-called 'Black Triangle' on the borders between southeastern Germany, Poland and the then Czechoslovakia.

Given the voluntary, indeed almost spontaneous, nature of the Euroregions, are they in a position to solve the wide variety of problems for which they have been set up? It may well be that the constitution of a Euroregion as a political–administrative entity does not of itself create an economically and socially integrated trans-border region (Krätke, 1996, p. 652). But, assuming that this is one of its aims, does the organization actually have the competence in the first place to implement change in such a way as to create an integrated region?

The effect of national administrative structures on inter-regional cooperation

The major hurdle to be overcome in any trans-boundary cooperative venture is the juxtaposition of two or more different national administrative structures, which will have an impact on decision-making ability. Some trans-national structures between the countries of Europe, which themselves exhibit varying degrees of centralism, fit together more easily than others.

The fact that the western half of Germany, for instance, has had a federal structure since World War II helped establish Euregio on the border with the Netherlands. Rather than a powerful central government, Euregio has been faced with only *Land* authorities in Germany and municipal authorities, in the region around Enschede, in the Netherlands. Success here may also have had something to do with the fact that this is the only such organization so far to set up a local trans-boundary parliament: the Euregio Council, created in 1978 and comprising 29 Dutch and 31 German representatives chosen by the municipalities. Even here, though, the Council, whilst it is 'an international private law organization', is only a 'consultative and coordinating body without a direct political mandate' and 'has no formal legislative or executive powers' (Scott, 1996, p. 93). It is, however, an organization which has grown up over a long period, has enormous support in the region, and operates as a very efficient lobby at senior government level. Thus, having started by organizing educational and cultural activities across the regional border, the participants are now able to work together on major projects to do with the economy, the environment, transportation and other issues.

The same is not true of trans-border relationships between Germany and France, although here such activity has just as long a history. Although over time cooperation has been successful, France's centralized administrative structure does not allow for a comparable integrated approach. This is the example most likely to be paralleled on Germany's eastern border with Poland. For post-Communist Poland has taken a long time to show any real decentralization of decision-making.

Nothing had been done in Poland since the then leader, Edward Gierek, removed the previous middle layer of regional administration (the *powiaty* – districts) in 1975 and increased the number of *województwa* (from 17 to 49) in order to simplify the lines of command of the Communist Party through to the *gminy* (local authorities). In early 1989 the first discussions took place between the Communist Party and representatives of *Solidarnosc* on the terms for the transfer of power. In early 1990 the Local Self-Government Act was passed and in May 1990 the first post-war democratic local elections took place. Nevertheless the real responsibilities of the new 'autonomous' local bodies (*gminy*) were at that stage by no means clear. What is certain is that the term 'decentralization' of government responsibilities is misleading and needs considerable qualification in this case. What devolution there was, has been limited in both decision-making powers and associated financial support to enable services to be delivered at local level (Zsamboki and Bell, 1997, pp. 177–8). Almost a decade of proposals, discussion and procrastination elapsed before agreement could be reached on a new regional administrative reform. Finally on 5 June 1998 decrees were published detailing the tasks and duties both of the 372 new *powiaty* and of the 16 new *województwa*.[5] On 11 October 1998 elections were finally held throughout Poland to set up new councils on a local, district and regional level, for implementation on 1 January 1999.

Throughout the 1990s, attempts within the democratization process to rearrange regional administrative structures in favour of some decentralization of decision-making have been frustrated by changes of government coalition – each with a short-term agenda rather than a commitment to longer-term reform. The apparent inability of the centre to reform itself has thus been seen as a major obstacle to democratic progress. Experts in regional policy watched a gradual *recentralization* taking place over some years. This did not help regional actors to adjust successfully to their new responsibilities and the accountability which is part and parcel of a participatory democracy (Regulska, 1997, p. 188; Gorzelak, 1992, p. 483).

The 1990 Act had nonetheless brought new agents of change onto the political and economic scene of public life. The *gminy*, Polish private entrepreneurs, foreign investors and non-governmental organizations created, with the help of new methods of training, a new and dynamic environment. Nevertheless, the *gminy* had little control over their tax revenue and were subject to a great deal of interference in their activities from the *województwa*, which are, after all, administrative units of central government. Local authorities therefore lacked clear rules and there was no political consensus on a division of power and responsibilities between local and central government (Grochowski, 1997, p. 216).

Thanks to the antipathy of the Peasants' Party (PSL) towards any reform, the first serious attempt to re-create an intermediary level of regional administration (*powiat*), intended to eliminate the political vacuum between parliament and the *gmina* was sidelined after the advent of the Pawlak government in 1993. Many regional officials in the PSL remained in post in the *województwa* from before 1989, and would very likely have lost their positions and privileges in the event of any major administrative reform enlarging the regional units (Paradowska, 1997, p. 20). Although the coalition partner, the Democratic Left Alliance (SLD), issued a position statement in support of decentralization, self-government and a 'citizen's society', this was generally thought to be simply part of a political campaign in advance of the next election (Regulska, 1997, p. 199).

On 1 January 1999, however, the regional map of Poland was transformed. There are now 16 new *województwa*, 372 *powiaty* and 2,489 *gminy*. The 16 new regions are responsible for regional administration. The potentially very powerful leader, the *wojewoda,* remains the regional representative of the Council of Ministers, so that the centralist ethic still remains. The *powiaty* have their own budget, though much of it consists of subsidies and grants from the state budget, albeit collected by the *powiat*. The *powiaty* have taken over responsibility for, amongst other things, public education, health and social services, transport and public highways, culture and tourism, water supplies, etc. But they are also ultimately answerable to the Council of Ministers. Technically speaking, the *gminy* have lost some of their previous competences to the district; but since they were never adequately funded, the difference in their role will probably not be very great.

It remains to be seen whether this new administrative reform brings in a genuine decentralization of powers to the regions, or whether it will be another masked *re*concentration of power by the centre. It will be important for the Euroregions that the Polish partners, which now include the new districts, can take independent decisions in their forward planning, without the strings being pulled by Warsaw as hitherto. The role of the new and bigger *województwa* will be crucial, possibly making decisions more complicated (Euroreport, 1998, p. 6).

The approach to regional planning, and specifically spatial planning, in Germany is, of course, very different, which only serves to emphasize the difficulty of achieving real cooperation where there is asymmetry between two administrative systems. Article 30 of the German constitution, the Basic Law, states that the execution of state tasks not otherwise allocated elsewhere in the Basic Law are the responsibility of the *Länder (Sache der Länder)*. Thus the *Länder* have the basic responsibility for spatial planning, although central government (the *Bund*) has the right to issue a legislative framework – and indeed did this in the Spatial Planning Law (*Raumordnungsgesetz*) of April 1965 (Scherer, 1998, p. 83). Individual *Land* legislation determines the organizational planning hierarchy, which will therefore differ from *Land* to *Land*.

Article 28 Para. 2 of the Basic Law bestows the right of self-government on the local authorities (*Gemeinden*), and also on local authority associations (*Gemeindeverbände*). Most importantly, the local authorities have an obligation to formulate a development plan (*Bauleitplan*), though the district (*Kreis*) is also involved since its district development plan contributes to the planning process at local authority level (Scherer, 1998, p. 98). In some *Länder* an intermediate level is also provided by regional planning associations, which include *Gemeinden* and *Gemeindeverbände*, plus possibly selected independent regional representatives. These associations exist for purposes of coordination; so that, for instance, local authority development plans, especially land-use planning, can be dovetailed without any loss of competence on the part of the local authority (Scherer, 1998, pp. 96–7).

It is clear that the two administrative regimes have virtually nothing in common: the Polish system hitherto has been highly centralized, with the regional and local actors rarely able to take independent decisions. This situation should improve with the advent of regional government reform, but will certainly not involve decentralization along German lines. The German system is highly decentralized, allowing for real participation in decision-making at both regional and local level. Any attempt to coordinate planning in the border region was therefore always likely to be difficult. Yet, by making use of EU funding mechanisms, the Euroregions have sought to act as a catalyst of change in the border region.

The limitations of EU funding mechanisms in trans-boundary planning

Both sides of this border have been, and still are, subject to deep structural transformation, and the Euroregions have attempted to address these problems by taking advantage of EU funding mechanisms. These were specially set up to mitigate the inherent problems of border regions and to meet the needs of the former Communist countries to modernize, in particular by creating employment. Fostering cross-national networks can also be seen as part of the European Commission's strategy to provide an important channel for the emergence of a European civil society and an alternative form of politics (Laffan, 1996, p. 97). Thus the eastern border region of the EU from Finland to Greece has been able to take advantage of two kinds of funding.

First, there is the Community Initiative INTERREG, originally set up in 1990 as an offshoot of the Structural Funds. By offering funding, with matching finance from sponsors in the regions concerned, INTERREG aimed to promote trans-boundary cooperation across both internal and external borders of the EU and to reduce the negative effects of isolation, both for the respective national contexts and for the EU as a whole. Refinements of the programme, INTERREG II and INTERREG IIC, were introduced in 1994 and 1996 to expand the number of areas of development which may benefit from this financial instrument. In INTERREG IIC, apart from introducing support for areas affected by drought and floods, the intention was to expand the planning unit enormously by supporting the joint development of 'transnational groupings going beyond simple cross-border cooperation and forming groupings involving at least three States (at least two of which are Member States), taking account of the size of possible territories involved in cooperation' (*OJ* C200, 1996, p. 24). The eastern German border region is set to receive over 400 million *ecu* in the current funding period (1994–99), partly due to the fact that all five of the new *Länder*, alongside regions such as Merseyside, southern Italy and Portugal, fall into the *Objective 1* category – that is, regions lagging behind in industrial development.

However, these funds can only be used by, and for, EU member states. So where the latter are cooperating with the former Communist countries, the second funding mechanism comes into play; that is, a special allocation of 'PHARE'[6] funding, known as PHARE-Cross-Border Cooperation or 'PHARE-CBC'. As the name implies, this funding is available specifically for cooperation across borders with EU member states. A sum of ecu 2.4 billion has been set aside for this purpose for the period 1995–99, and of the 150 million distributed in 1995, 90 million were allocated to Poland.

Applications for INTERREG funding for cooperation across the EU external border have to come from the member state concerned in the form of an Operational Programme 'covering, where appropriate and practicable, measures on both sides of the border and indicating the measures or parts of measures for which assistance under the Structural Fund regulations is requested' (*OJ* C180, 1994, p. 65). An overall framework for such trans-border planning is established by the Intergovernmental Commission. The individual cross-border projects which make up each Operational Programme are proposed by local authorities and other organizations, private or public, discussed in the Euroregion, and then submitted for approval to the competent authorities on either side of the border.

As earlier discussion has anticipated, applications for INTERREG and PHARE funding throw into relief considerable differences between German and Polish practice. In the case of Germany, Operational Programmes are submitted by the *Länder*, since they are the competent authority for regional and spatial planning. In Poland, project applications for PHARE monies are coordinated by several ministries of the central government, with a particularly important role played by a special office responsible for trans-border matters. This means that not only is there a distinct mismatch between the decision-making authorities in the two countries, but also that the arrangements generate a level of bureaucracy which is not helpful to the task in hand.

There are, then, a number of problems associated with this process. First, the two separate funding mechanisms require separate applications from the initial submission to the national authority, right through to the ultimate decision-making bodies in Brussels. Thus, although cross-border projects are discussed in the respective region as a whole, there is no guarantee that they will be submitted in the agreed form by both sides – especially bearing in mind the asymmetrical decision-making structures discussed above. One particularly unfortunate example of what can go wrong as a result was provided by the Euroregion *Pro Europa Viadrina* in 1997. Being very flat, the Polish part of this central area of the border region was disastrously affected by the flooding of the river Oder in July of that year, to the extent that large parts of Slubice, opposite Frankfurt (Oder) had to be evacuated. The Euroregion had proposed emergency arrangements for just such an eventuality, but it turned out that the Polish authorities had not yet signed the relevant documentation to agree to such arrangements. The German emergency services had no legal basis for crossing the river to provide assistance.

A number of proposals for cross-border cooperative measures in this and other Euroregions, such as a sewage disposal works in Guben/Gubin in the Euroregion *Spree–Neisse–Bober*, and a number of tourism projects in all the regions, have indeed been successful. But frustration often prevails. A major source of frustration is the time taken to make decisions about whether the appropriateness of projects submitted on either side. Projects go from the Euroregions to their respective regional or national authorities as part of the development plan for their own boundary region. The projects are then subjected to lengthy, not to say laborious, evaluation at ministerial level in both Germany and Poland.

Within *Land* governments, different ministries are responsible for different areas of activity, so that disagreements amongst ministries can delay matters considerably. One particular source of dissent has been an imbalance between the large number of projects on social and environmental issues, by comparison with business-related and economic projects. This probably has to do with regions being as yet underdeveloped in economic terms. The newly market-oriented business communities of eastern Germany and Poland are not yet ready to trust and cooperate with potential partners on the other side of the border. Even though they emerged from the same camp in 1989, at present the two sides undoubtedly see each other in two very different economic environments – thus continuing the sense of distinct, even rival, economic trajectories fostered during the Communist period.

At any rate, a number of initiatives have been blocked in the *Land* ministries due to disagreements at various stages. Other conflicts of interest arise from the fact that the Euroregions and their projects do not fit neatly into the *Land* boundaries. *Land* Brandenburg, for example, has responsibility for agreeing and submitting *competing* operational programmes from *Pro Europa Viadrina*, *Spree–Neisse–Bober* and part of *Pomerania*. On the Polish side too there are disagreements, mainly due to the fragmentation of decision-making between the *województwa* and central ministries and agencies in Warsaw, which render the *gminy* in the border region relatively powerless. Add to this the fact that the *Länder* and the *województwa* each guard their own prerogatives jealously, and there is ample room for posturing and grievances, all of which leads to delays (Kennard, 1997, p. 57).

It is no surprise to discover that there are further hold-ups in Brussels. Once the documentation finally reaches the Commission, the rest of the cumbersome process, which may include returning the Operational Programme for modifications, can take up to two years. This causes frustration amongst the regions concerned, and particularly amongst the project sponsors, who have to reserve adequate monies for the duration of the entire approval process, anything up to three or four years. To emphasize this point, the INTERREG II operational programmes for 1995–99 were not approved until July 1995, so that the monies for that year had to be redistributed over the remaining years. The EU funding mechanisms, imaginative and well-meant as they are, lead to clashes of interest. In sum, if considerable streamlining does not take place, both at a national and European level, there is a danger that local actors will lose heart and the whole process will lose momentum, just as Poland is moving closer to membership of the EU.

Prospects for the future of trans-national planning in the region

So what is to be concluded about the future of the German–Polish border region? To what extent and in what sense is it likely to remain an awkward margin in the European context? The geopolitical position of the region, with its considerable historical 'baggage', makes its problems perhaps at once the most difficult and the most important to solve of any border in Europe. And yet the events of the final decade of the twentieth century indicate that the erstwhile enemies on either side have decided to put the past behind them, to embrace the brave new Europe, and to construct a positive present and future.

Some political commentators point to an apparent difference in the overall reactions in Poland and eastern Germany to their newfound economic freedom. Baring speaks of 'stagnation and apathy' in the area of eastern Germany, whereas 'in Poland the spirit of enterprise never died' (1998, p. 21). 'East Germany had a command economy before communism, during it and afterwards', echoes Sanai (1998, p. 7). There is a palpable sense here of two *different* margins, each attached to their claim to belong to Europe, but with different models of what *is* European: the German interventionist model versus the Poles' age-old enterprising spirit. Contending senses of 'self' face different 'others'; but each claims membership of something else called 'Europe'. Currently, the differences are *heightened* by the EU/non-EU divide, since it lies between Poland and eastern Germany, but will, after the first wave of accession, shift to include Poland along with the other Central European states. The borderlands people may find it easier to adapt to that new situation than the two nations as a whole. If the border continues, with the help of the Euroregions, to take on the character of a 'frontier' rather than that of a 'boundary' – with the openness to, and acceptance of, the other side which that implies – then the situation will continue to look hopeful.

However, there are still various hurdles to be overcome and unknowns to be taken into account. First, will the economy of the region be able to develop as an integrated economic area? The 'bazaar economy' has grown very fast and demonstrates the possibility of a cross-border economy. But this is trade of a very specific and one-way kind (akin to the *maquiladora* model), which may become difficult to square with the priorities of INTERREG. As enormous queues of traffic form daily to cross the border, particularly at the Oder bridge at Frankfurt (Oder),[7] the practicalities of travelling may focus attention on the increasing inadequacies of the transport and communications systems. Since the Schengen Agreement, traffic on this external EU border has been subject to increased delays, due to tighter controls on immigration, asylum-seekers and smuggling.

How then can the regional economy proceed in these rather difficult circumstances? The proposals of the Euroregions are aimed at encouraging local cooperation, but the evidence so far is that relationships between local trans-border enterprises are still relatively rare. More frequent is 'supra-regional' investment, which overshoots the narrower border region, and relationships between yet more distant international enterprises (Krätke, 1996, p. 663). This does not augur well for the border region. Will the region become a backwater of Berlin, or a new 'tiger' economy, challenging European established economic centres? Certainly Berlin, and also Dresden, will continue to exert a pull on the German side; but on the Polish side there is evidence of increasing investment around the Szczecin metropolis and smaller centres such as Gorzów and Jelenia Góra.

Real cooperation will certainly be difficult without some cross-border regional planning mechanism which is binding. At this point in time the only vehicle for such cooperation is that provided by the Euroregions. The major problem remains, though, as they labour under three major institutional difficulties (Roch *et al.*, 1998, p. 56). First, the Euroregions do not have any legal or institutional competence either in the border region or in the respective national part-regions. Secondly, the competent authorities for planning coordination in both borderlands are situated outside these areas. Thirdly, they belong to two differently functioning administrative structures. The only solution is for the respective national authorities to think the unthinkable and have the courage to accept the concept of real trans-border responsibility for the purposes of regional planning.

There is not yet a thoroughgoing sense of regional identity in this part of Europe. Indeed, it would be surprising if the vicissitudes of history had been forgotten after such a few short years. But there is a discernible sense that there are problems to be solved which transcend the border and will benefit from a common approach. If there is the will to adapt the institutional and financial context appropriately, then regional identity could contribute to the development of a European identity in the wider sense. In that case, with EU enlargement, the marginal character of the region as a whole will disappear, and the frontier of Europe will shift elsewhere.

Notes

1. See *European Outline Convention on Transfrontier Cooperation between Territorial Communities of Authorities* (Madrid: Council of Europe, 1980).
2. *Maquiladora* = industrial assembly plants, set up on the US–Mexican border to import industrial components duty-free from the US into Mexico, where they are assembled by low-cost labour and exported back to the US with special tax advantages.
3. The basic conceptual documents describing the *raison d'être* of the Euroregions was published as *Entwicklungs- und Handlungskonzept* (1993–4), and as *Operationelles Programm* (1995). The German–Czech border region is covered by three further Euroregions, in addition to Neisse: Elbe/Labe, Erzgebirge and Egrensis.
4. *Województwo*, also known in English as 'voivodship'; this is a regional administration unit in Poland rather akin to the French *département*. From 1 January 1999 a new regional administrative reform was introduced. See below.
5. *Rzeczpospolita*, 5 June 1998.
6. PHARE = Poland–Hungary Aid and Reconstruction for Europe. This fund was initially set up by the European Community in 1990 to support Poland and Hungary in their efforts to reconstruct their economies along

market lines, but has since been extended to include all the former Eastern bloc countries in central Europe. A similar fund exists for the republics of the former Soviet Union: TACIS = Technical Assistance for the Commonwealth of Independent States and Georgia.

7 Approximately 50 per cent of all heavy goods road traffic to Poland and the CIS states crosses the border on the motorway at Frankfurt–Swiecko, and similar figures apply for the railway connection (Krätke, 1996, p. 655). There are also hours-long queues of private traffic on the bridge between Frankfurt and Slubice, the major East–West route between Berlin and Warsaw, extending west to Paris and east to Moscow.

References

Anderson, M. (1982) 'The Political Problems of Frontier Regions', *West European Politics*, **5**(4), 1–17.

Baring, A. (1998) 'Do czego potrzebna jest Polska?' (What use is Poland?), *Polityka*, **17** (25 April), 20–1.

Curzon, Lord, of Kedleston (1908) *Frontiers*, The Romanes Lecture, 1907, 2nd edn (Oxford: Clarendon Press).

Entwicklungs- und Handlungskonzept für die Euroregion Neisse-Nisa-Nysa (Friedrichshafen: Dornier GmbH, vol. 1, 1993; vol. 2, 1994).

Entwicklungs- und Handlungskonzept (Operationelles Programm) Europaregion Pomerania (Econometrika GmbH, 1993).

Entwicklungs- und Handlungskonzept für die Euroregion Pro Europa Viadrina (Berlin: Software Union, 1993).

Entwicklungs- und Handlungskonzept für die Euroregion Spree-Neisse-Bober (Friedrichshafen: Dornier GmbH, 1993).

Euroreport der Euroregion Pro Europa Viadrina, March 1998.

Gabbe, J. (1992) 'Grenzüberschreitende Netzwerke' (Transborder networks). *Stadt und Gemeinde* (Town and Municipality), **3**, 92–8.

Ganster, P. (1996) 'On the Road to Independence? The United States–Mexico Border Region', in J. Scott *et al.* (eds), *Border Regions in Functional Transition: European and North American Perspectives on Transboundary Interaction* (Berlin: Institute for Regional Development and Structural Planning – IRS), pp. 171–91.

Gorzelak, G. (1992) 'Polish Regionalism and Regionalization', in G. Gorzelak and A. Kuklinski (eds), *Dilemmas of Regional Policies in Eastern and Central Europe* (Warsaw: University of Warsaw, European Institute for Regional and Local Development), pp. 465–88.

Grochowski, M. (1997) 'Public Administration Reform: an Incentive for Local Transformation?', *Environment and Planning C: Government and Policy*, **15**, 209–18.

Gruchmann, B. and Walk, F. (1996) 'Transboundary Cooperation in the Polish–German Border Region', in J. Scott *et al.* (eds), *Border Regions in Functional Transition: European and North American Perspectives on Transboundary Interaction* (Berlin: Institute for Regional Development and Structural Planning – IRS), pp. 129–38.

Jones, S. B. (1959) 'Boundary Concepts in the Setting of Place and Time',

Annals of the Association of American Geographers, **49**(3), 241–55.
Kennard, A. (1996) 'Issues of Identity in the Polish–German Border Region', *Journal of Area Studies*, **8**, 106–19.
Kennard, A. (1997) 'A Perspective on German–Polish Cross-Border Cooperation and European Integration', in M. Anderson and E. Bort (eds), *Schengen and EU Enlargement: Security and Cooperation at the Eastern Frontier of the European Union* (Edinburgh: University of Edinburgh, International Social Sciences Institute), pp. 53–61.
Krätke, S. (1996) 'Where East Meets West: the German–Polish Border Region in Transformation', *European Planning Studies*, **4**(6), 647–69.
Krätke, S. (1997) 'Regional Integration or Fragmentation? The German–Polish Border Region in a New Europe'. Paper presented at the EURRN conference on 'Regional Frontiers', Frankfurt (Oder), Germany, 20–23 September.
Kristof, L. K. D. (1959) 'The Nature of Frontiers and Boundaries', *Annals of the Association of American Geographers*, **49**(3), 269–82.
Laffan, B. (1996) 'The Politics of Identity and Political Order in Europe', *Journal of Common Market Studies*, **34**(1), 81–102.
Macków, J. (1995) 'Die Normalisierung der neuen alten Nachbarschaft: Zum aktuellen Stand der deutsch–polnischen Beziehungen', *Aus Politik und Zeitgeschichte*, **39**(95), 32–9.
O'Dowd, L. and Wilson, T. M. (1996) *Borders, Nations and States* (Aldershot: Avebury).
Official Journal of the European Communities (1994) No. C180 (INTERREG II), 1 July, pp. 60–8.
Official Journal of the European Communities (1996) No. C200 (INTERREG IIC), 10 July, pp. 23–8.
Operationelles Programm des Landes Brandenburg für die EG-Gemeinschaftsinitiative INTERREG II für die Euroregionen Spree-Neisse-Bober und Pro Europa Viadrina in der brandenburgisch-polnischen Grenzregion (Potsdam: Land Brandenburg, 1995).
Paradowska, J. (1997) 'Powiat zatupany' (Districts in trouble), *Polityka*, **5** (1 February), pp. 20–2.
Regulska, J. (1997) 'Decentralization or (Re)centralization: Struggle for Political Power in Poland', *Environment and Planning C: Government and Policy*, **15**, 187–207.
Roch, I., Scott, J. W. and Ziegler, A. (1998) *Umweltgerechte Entwicklung von Grenzregionen durch kooperatives Handeln* (Environmentally Friendly Development of Border Regions via Cooperative Action) (Dresden: IÖR-Schriften 24, Institut für ökologische Raumentwicklung e.V.).
Rzeczpospolita, 5 June 1998.
Sanai, D. (1998) 'Bridging the Oder gap', *The European*, 4–10 May, p. 7.
Scherer, B. (1998) *Regionale Entwicklungspolitik: Konzeption einer dezentralisierten und integrierten Regionalpolitik* (Frankfurt am Main: Peter Lang).
Scott, J. W. (1996) 'Dutch–German Euroregions: A Model for Transboundary Cooperation?', in J. Scott *et al.* (eds), *Border Regions in Functional Transition: European and North American Perspectives on Transboundary Interaction* (Berlin: Institute for Regional Development and Structural Planning – IRS), pp. 83–103.

Stryjakiewicz, T. and Kaczmarek, T. (1997) 'Transborder cooperation and development in the conditions of great economic disparities: the case of the Polish–German border region'. Paper presented at the EURRN conference on 'Regional Frontiers', Frankfurt (Oder), Germany, 20–23 September.
Zsamboki, K. and Bell, M. (1997) 'Local Self-government in Central and Eastern Europe', *Environment and Planning C: Government and Policy*, **15**, 177–86.

Part III

The Political Possibilities of Marginality

8

EU Territorial Governance and National Politics: Reshaping Marginality in Spain and Portugal

José Magone

Introduction

For decades Portugal and Spain were isolated from democratic development in other European countries by authoritarian regimes based on a moderate nationalist ideology and intent on preserving traditional ways of life. Industrialization and modernization in the 1960s and 1970s eroded the corporatist and nationalist ideological foundations of these regimes. As the rift between authoritarianism and an emerging culture of alienation grew, isolation from Europe became less possible in the mid-1970s (*Os Portugueses*, 1973; Pina and Aranguren, 1976). Tourism, mass media and emigration questioned monolithic definitions of identity. Nationalism became obsolete in a time when the *Zeitgeist* (Linz, 1978, p. 6) was dominated by Socialist ideologies which encouraged the transition to democracy in both countries.

The Portuguese Revolution of 25 April 1974 can be seen as a relatively spontaneous explosion of emotions; an outburst of participation and experimentation in reaction to 48 years of dictatorship and the long night of fascism (Figueiredo, 1975; Ferreira, 1983). In contrast, Spain's transition to democracy, although sustained by the political mobilization of the working classes, was led by the old and new political elites (Maravall, 1982, p. 13). But in both cases, constitutional settlements were crucial both in the establishment of the new democratic regimes, and in influencing the nature and direction of democratization in Spanish and Portuguese politics. The Portuguese constitution emphasized participatory democracy within the territorial frame of reference of the unitary state. The Spanish constitution was significant as much for its prescription regarding the new regime as its arrangements for a new centre–periphery settlement (Bonime-Blanc, 1987). In both cases implementation of the constitution over the past two decades has led to considerable spill-over effects. In Spain in particular regionalization and regionalism have re-created democratic politics.

As well as transforming and liberating Iberian societies, democratization brought the two countries closer to one another and to the other countries of Western Europe, and ultimately led to the accession to the European Community in 1986. Spain spread its linkages to the outside world, playing a more active role in international relations. Spanish external relations reinforced the process of democratization, and simultaneously transferred the experience of democratic transition to other countries, particularly in Latin and Central America (Gillespie, 1995, p. 196). Although slower in establishing a presence in a rapidly changing international environment, the Portuguese presidency of the European Community in 1992 confirmed the arrival of Portugal on the international stage (Magone, 1997, pp. 164–70). Iberia could no longer be considered marginal in Europe.

This chapter examines the transformation of the Iberian countries in the new Europe in the light of developments since the Maastricht Treaty and the ratification of the Amsterdam Treaty. The first section deals with the current state of governance in the European Union. The second section explores the new emerging territorial politics of the European Union, which are reshaping the boundaries between supranational, national and subnational levels. The following sections examine the corresponding reshaping of boundaries and levels of governance in the emerging civil societies in Spain and Portugal. The third section focuses on the relationship between the centre and the regions in Spain, while the fourth section looks at populist nationalism and regionalization in Portugal in relation to the current debate about Europeanization. European integration in general, and the territorial policies of the European Union in particular, have been a major factor in reinforcing regionalization and democratization processes in Iberia. These processes in turn have irrevocably altered perception both from within and beyond Iberia. Their place in the regionalization of Europe has made Spain and Portugal no longer 'marginal' in a pejorative sense. They now occupy positions firmly at the core of the political development in the European Union.

The problem of mismatching interfaces in the governance of the European Union

The most recent heuristic theoretical devices seem to concur that the EU is moving towards a multi-level framework of decision-making comprising the supranational, national and subnational levels. The 'Europe of the Regions', the 'Europe of the Nations', the 'Europe of the Citizens', the 'Europe of the Multinationals' clearly indicate new discursive divides in the Euro-Polity. From amongst these, the importance of the regions in reshaping the Euro-Polity in the 1990s cannot be stressed enough. The transformation of regional policies at European level has led to a growing involvement of the regions in the supranational framework (Marks, 1993, pp. 391–410; Hooghe and Marks, 1996, pp. 73–91). The best example is the Committee of the Regions and Local Authorities (CoR). Despite its lack of decision-making competences, it is of considerable symbolic importance to subnational actors.

With Maastricht, however, several new institutions were created that have still to achieve a level of consistency and congruence with the other institutions. For the time being inter-institutional inconsistencies weaken the efficiency of decision-making at European Union level. Administrative culture within the EU lacks an overarching rationale, and is fragmented and dependent on the nature of the particular institutions. The Council and the Commission are dominated by a myriad of intergovernmental committees dealing with policy formulation while, on the other side, the European Court of Justice has gained a considerable autonomy in relation to the member-states. The political system '*sui generis*' is still in flux. Its final outcome will be a new kind of polity, resembling neither the nation-state, nor a federation. Rather it will institutionalize the pendulum swing between ambitious future-oriented supranational policy-making initiated by the Commission, and present- or even past-oriented policy-making of member states (Schmitter, 1996a, pp. 137–8). The final outcome, highly dependent on the voluntarism of political actors at the European level, will be a compromise between twin processes of integration and disintegration (Tranholm-Mikkelsen, 1991).

Recent debate trying to capture the nature of governance in the European Union has been confusing and difficult to assess. Constant and rapid evolution of structures of opportunity and institutional frameworks creates major difficulties in the definition of the EU's governance. Helen Wallace defines it as a policy pendulum between national and supranational solutions. While the national side includes dissonant national policies and diverse ideas in country-level polities, the supranational comprises congruent national policies and shared ideas in a partial European polity. Both levels tend to converge in response to the challenges of globalization (Wallace, 1996, pp. 12–13). EU governance and processes of integration are actor-centred and dominated by negotiation (Marks, 1996). The consequent voluntarism of political, social and economic actors in the European Union is contributing to a self-generating process of structuration of the Euro-Polity, simultaneously differentiating these structures, and enhancing their autonomy with the passing of time. This actor-centred, but also structural development of European integration is accompanied by a growing consistency and integration in the policy process. Thus, for example, looking at the politics around the present budget structure, one can clearly identify a growing change from passive pay-out policies to pro-active ones of social, economic and political transformation.

Philippe Schmitter has made a thorough step-by-step investigation of the nature of EU governance. He defines governance in modern democratic political systems as an

> ensemble of patterns that determines (1) the forms and channels of access to principal governmental positions; (2) the characteristics of the actors who are admitted to or excluded from such access; (3) the resources or strategies that these actors can use to gain access; and (4) the rules that are followed in the making of publicly binding decisions. To produce its effect, the ensemble must be institutionalized, i.e. the various patterns must be habitually known, practised and accepted by most, if not all, of the actors. Increasingly, this has involved their explicit legalization or constitutionalization, but many very stable regime norms can have an implicit, informal, prudential or precedential basis.
>
> (Schmitter, 1996b, pp. 2–3)

According to this definition, based very much on actor-centred structural considerations the EU is as yet only a system of governance in the making. From the different parts of his definition one can recognize the main problem of the Euro-Polity; the incompleteness and the blurred, undefined nature of the 'ensemble of patterns'. Supranational, national and subnational 'ensembles of patterns' are *mismatching*. Only partially linked, they lack an overarching rationale. The Euro-Polity remains undefined even as a political system '*sui generis*' (Hrbek, 1989, pp. 97–8). Its partial completion is a general obstacle to the establishment of an overarching rationale connecting the presently mismatching 'ensemble of patterns' between the different levels, particularly the national and supranational. The idea guiding this chapter is that the EU's accommodation with marginality in Spain and Portugal may illustrate how it can overcome the problem of mismatches. The development of actor-centred, but also structured, governance within the multiple levels of regional policy is the proving ground for this claim.

The territorial policies of the European Union

The transformation of the rationale of governance

The structural and regional policies of the European Union can be considered as *supranational* territorial policies designed in congruence with *national* territorial policies. The approval of Common Support Frameworks since 1988 for the different countries reflects the logic of EU territorial policies. The central rationale of these territorial policies is to achieve social cohesion across EU territory, as a counterpart to the enhanced competition resulting from the Single European Market (SEM). The poorest regions of the Community are eligible for compensatory payments. In EU territorial policies the 'region' became a central category for the eligibility to structural funds. It became an actor in its own right with the ability, in partnership with the national government and the Commission, to shape the priorities of development for a determined period of time (1989–93, 1993–99).

Until the mid-1970s the nation-states were the major actors of the European integration process. The call for a return of 'Europe des Patries' by Charles de Gaulle in 1965 merely re-asserted the supremacy of the nation-state in the process of European integration. Plans to complete the single market as proposed in the European Economic Community Treaty had to wait almost twenty years to be relaunched.

But then accessions to the EC intervened: the United Kingdom, Ireland and Denmark in 1973; Greece in 1981; Portugal and Spain in 1986 and Austria, Sweden and Finland in 1995. Within two decades the European Community/European Union had become more heterogeneous, particularly in relation to the economic divergences between peripheral areas and the core countries (Benelux, Germany and France). With the accession of Greece in 1981 and Spain and Portugal in 1986 the number of regions 'lagging behind' within the Community's territory increased considerably. Enlargement brought a fresh plurality of voices, emphasizing different aspects of the deepening process. Furthermore, the enlarged EC/EU had to seek ways of reconciling aspects of its own development: to reconcile, on the one hand, convergence between the peripheral and core member-states and, on the other hand, the pursuit of the SEM based on the principle of competition (Belloni, 1994). Although there seems no question that economic and social cohesion was an important pre-condition to a more equitable SEM based on fair competition, it is still not clear how 'economic and social convergence' between peripheral areas (Portugal, Greece, part of Spain, part of Italy and Ireland) and core ones (Germany, United Kingdom, France, Benelux, Finland, Sweden, Austria) can be measured. The tendency to present a dichotomous picture between rich and poor countries disguises the fact that in the 1980s and 1990s some regions economically lagging behind have done better than others in converging to the core countries. Yet, that said, the territorial policy was intended to address the problem of mismatches by identifying a new EU-wide, overarching rationale for governance.

The reform of the structural funds in 1988: the fusion of European and national policies

The reform of the structural funds in 1988 first upgraded the role of the regions in the framework of European regional policy. The new CoR gave consultation rights to the subnational actors and established that decision-making should be based on the principles of partnership and subsidiarity and a more dialogical relationship between the Commission and the nation-states. After the structural funds reform of 1988, the allocation of financial transfers has been decided in conjunction with the member states. In general, a Regional Development Plan has to be submitted to the Commission. The Commission then drafts a Common Support Framework (CSF) based on priorities agreed by the member states in the Regional Development Plan. The CSF specifies the allocation of funds to the different categories in the Regional Development Plan. Under this process of financial allocation the nation-state remains the 'gatekeeper' of the structural funds, although the regions are involved.

Problems associated with the growing complexity of the Community structures in relation to allocation, implementation and monitoring of Community law and policy can be eased by the integration of the subnational, regional actors in the decision-making process. For example, the conference on INTERREG in December 1992 brought together several representatives of the regions concerned. The experiences reported in conference contributed to the Commission's drafting of INTERREG II. The creation of networks is transforming the relationship between the different levels of decision-making and encouraging a vertical integration in the European political system (Engel, 1993, p. 94).

The CoR – with its consultative powers related to education policy, vocational training, culture, public health and trans-European networks – was already a crucial element in the new European architecture. Of special importance have been its consultative powers related to the European Regional Development Fund and other structural funds, and the right to be informed on a wide range of community policies from common agricultural policy to harmonization of national legislations related to the internal market, to education and industry. Nevertheless, the CoR is subordinate to the Economic and Social Committee, and the heterogeneity of its membership (coopted in a variety of ways) is a major impediment to the achievement of coherence (Kalbfleischer-Kottsieper, 1993, pp. 137–8; Loughlin, 1996).

The thickening of networks of *sub*national actors at the *supra*national level has been complemented by an increase in the number of offices set up by regional governments lobbying the European Commission. Representation of the different regions is uneven, reflecting different national traditions and territorial structure, but there are some 54 offices of regions and local authorities coming from six countries (Belgium, Germany, UK, Spain, Italy and France), representing three sevenths of the population (Loughlin, 1996, pp. 147–8). Spanish regions are very well represented: in 1993 nine regions had offices in Brussels (Andalucia, Catalonia, Galicia, Madrid, Extremadura, Pais Vasco, Comunidad Valenciana, Murcia and Canarias). The main function of representation at supranational level is to exploit multiple points of access. Offices have a limited ability to influence policy-making, but can obtain early warning of policy initiatives entering the pipeline, minimizing the hazards of a highly unpredictable environment (Loughlin, 1996, p. 58).

All these sets of actors complement the official vertical structure of European regional policy. Notwithstanding that the importance of European regional policy has recently been downplayed by leading experts on the subject, one still has to see the latest developments as a qualitative leap in relation to the pre-1988 situation.

The fifth report on the socioeconomic situation and development of the regions

The fifth period report on the social and economic situation and development of the regions in the community, published by the European Commission in 1994 under the title 'Competitiveness and Cohesion: Trends in the Regions', developed a long-term vision of regional policy as an instrument to encourage even development. It integrated regional policy as an instrument to create even development across the territory. Regions lagging behind *Objective 1* after the reform of the structural funds would receive 73 per cent of funds in 1999 (in comparison to 63 per cent in 1989 and 65 per cent in 1993), the share of the four poorest member states would rise from 50 per cent in 1993 to 54 per cent in 1994. The population covered had risen from 43 per cent in 1989 to 52 per cent in 1993, Portugal, Greece and Ireland covering the whole territory (European Commission, 1995, p. 128). Specific programmes tailored to the needs of certain regions or sectors reinforced the principle of economic and social cohesion (European Commission, 1995, p. 132; European Commission, 1994).

The report also included a chapter on the adjustment of the structures of the nation-states to the new regional policies of the European Union. The chapter deals with different approaches to regional policy in the late 1980s and 1990s. In general, it comes to the conclusion that, especially in northern countries, comprehensive regional policies are declining in favour of selective regional incentive ones. Southern European countries, by contrast, have experienced a boom in regional policies focusing on provision of infrastructure (European Commission, 1995, pp. 135–42). As EU-wide regional policy increasingly addressed territory-wide development, then, national policies diverged in character and effect.

Agenda 2000: a redefinition of the structural policies

On 16 July 1997 the Commission presented a report called *Agenda 2000*, which addressed the financing of the European Union in view of the forthcoming enlargement to the east. Though all the financial aspects of *Agenda 2000* remain provisional, a reduction of regional funds for the southern European countries appears likely. Indeed, the incorporation of the first candidates for accession (Cyprus, Poland, Hungary, Czech Republic, Slovenia and Estonia) will probably lead to major institutional changes in the European Union. While the EU will be making available ecus 275 billion for the regional funds, ecus 230 billion of this will be reserved for the existing fifteen member countries, and other financial programmes will be offered to eastern applicant countries during the enlargement period. According to *Agenda 2000*, only the less developed regions will be entitled to receive structural funding. In other words, the rationale of the structural funds will be ever more tailored to reduce the gap between the richest and very poorest regions. The share of the population eligible for structural funds will decline from 51 to 40 or perhaps 35 per cent. At the same time, the expenditure for the Common Agricultural Policy will be reduced considerably. Spain and Portugal will be considerably affected by these changes (Jochimsen, 1998; Goybet, 1997). In short, between 1988 and the end of the century we can observe EU territorial policy gradually defining a new cohesion rationale for EU-wide governance.

Transformation at the margin: Spain

The process of regionalization in Spain began after the adoption of the constitution in 1978, and gathered pace after the accession to the European Community. A closer relationship to Europe increased the influence of the regions in Spanish polity. Europeanization and regionalization also constituted complementary means to further democratize Spanish political structures.

Regionalization in Spain

The Spanish subnational level consists of regional, provincial and local governments. The regional level, the 'autonomous communities', are regarded as coordinating tiers between the provincial and the local. Involvement of the regions in constitutional negotiations resulted in '*el estado de autonomias*' (the state of autonomies), a compromise which at one and the same time acknowledged the state as unitary authority and granted autonomy to certain historical regions, such as the Basque Country and Catalonia. Regionalist parties from these regions were involved in the negotiation of the settlement. The final formula, though ill-defined and open-ended, resulted in the transformation of a highly centralized state into a semi-federal structure.

Between December 1979 and August 1983 seventeen autonomous communities came into existence. In the face of this regionalist dynamic, which it felt was getting out of hand, the main party of transition, the government of the Union del Centro Democratico (UCD), attempted to impose uniformity and governmental control over the process. The Organic Law for the Harmonization of the Autonomy Process (*Ley Organica para la Armonizaciôn del Proceso Autonomico – LOAPA*) enacted by the Cortes in June 1982, was opposed by the existing regions, particularly the Basque Country and Catalonia.

In the event the Constitutional Court annulled LOAPA and so strengthened the position of the Spanish regions (Graham, 1984, pp. 262–5). Although the Socialist government after 1982 had several conflicts over the transfer of competences with regions such as the Basque Country and Catalonia, Spain experienced a steady devolution of powers to the regions in the second half of the 1980s. In 1985 over 300,000 civil servants were allocated to the regions. By 1992 approximately 432,186 posts had been transferred from central to regional government (Heywood, 1995, p. 156).

Uneven progress towards statutes of autonomy led to different levels of devolution. The historic Basque Country and Catalonia regions were able to achieve a greater range of competences than regions such as Extremadura or Asturias. Nevertheless the dynamic of regionalism has shifted the overall boundaries between central and regional government. In the 1990s regional political actors gained prominence at national level, prompting discussion of restructuration of the second Chamber of Parliament, the Senate, into a territorial chamber, to strengthen the representation of the regions at national level. In parallel, the impact of the European Union in Spain led to the emergence of the model similar to the 'Europe of the Regions'. Spaniards are very active in the CoR: for example, Jordi Pujol, the president of the Catalonian government, or *Generalitat*, and Pascual Maragall, the mayor of Barcelona. The constant discussion over the Inter-territorial Compensation Fund between the regions and central government is increasingly transforming Spain into a fully federalist state.

Constitutional provisions for the achievement of regional autonomy have created a four-tier division of regions. The historic regions of the Basque Country and Catalonia, which were already recognized during the Second Republic, enjoy the highest level of autonomy. In the second tier are the Galicians and Andalucians at almost the same level as the historic regions. Valencia, Navarre and the Canaries have achieved autonomy, with a lower level of powers, and the other ten autonomous communities – Castilla y Leon, Castilla La Mancha, Extremadura, the Balearic islands, Madrid, Aragon, Asturias, Cantabria, Murcia and Rioja – have the lowest level of granted competences (Solé-Villanova, 1989, pp. 209–13; Brassloff, 1991, pp. 57–68). Yet the four-tier division is under constant review as boundaries between central and regional levels shift continuously towards the latter. During 1995 six regional parliaments, including Asturias, Cantabria and Castilla-La Manc, decided to reform their statutes. Aragon demanded a special regime of finance such as exists for the Basque Country and Navarre. All these reforms of statutes were then discussed by the Cortes in Madrid, without a decision being reached. Yet disputes over competence between Central and Regional government have declined considerably in the past two or three years. The growing self-confidence of regions, their recent impact on the national political system and the development of multiple identities has made Spain an interesting paradigm for other countries (*El Pais, Anuario*, 1996, pp. 122–3).

At the moment one-third of the national budget is spent at regional and local levels. Nevertheless, the allocation of resources is still very highly centralized. Only the Basque Country and Navarre have the right to decide tax matters for themselves. Other regions are allocated portions of the tax-share from central government. The seven autonomous communities with a high level of competence are able to take part in the decision-making process related to education and health. The competences of Catalonia and the Basque Country extend to policing. All autonomous communities are subject to the common financing regime, with the exception of the Basque Country and Navarre – due to their historical *fueros* (charter rights). Some autonomous communities such as Asturias, Cantabria, Madrid, Murcia, La Rioja and Navarre are allowed to levy separate provincial revenues. Moreover, regional taxes can be levied by six regions (Andalucia, Canary Islands, Catalonia, Murcia, Valencia and Basque Country). Seven autonomous communities have a high level of responsibilities particularly in the areas of health and education. The other ten autonomous communities have only a low level of powers (Aragon, Balearic Islands, Cantabria, Castilla y Leon, Castilla, La Mancha, Extremadura, Madrid, Murcia, Rioja).

The Inter-territorial Compensation Fund (*Fondo de Compensacion Interterritorial* – FCI) was devised to transfer funds from the richer regions to the less advantaged ones such as Andalucia and Extremadura. It redistributed over £7.5 billion between 1984 and 1991. Yet despite this compensating mechanism, Spanish regions are still underfunded. Most of the 8,056 local authorities are similarly underfunded and have difficulty coping with the tasks allocated to them. The financial grants to local authorities are transferred from the central government via the intermediation of the regional governments. In general terms it is felt by the President of the Spanish Federation of Spanish Municipalities and Provinces, Rita Barbera, that the central state must decentralize even further the allocation of resources according to a key of 50/25/25, that is: 25 per cent of resources to be devolved to the autonomous communities and 25 per cent to the local authorities (Barbera, 1996, p. 123). A major problem for local government is the lack of cooperation between different municipalities, and their sheer number. Over 7,000 municipalities have less than 5,000 inhabitants, necessitating a considerable dispersal of resources.

An additional, intermediary tier between regional and local government exists in the province, a tier established during the nineteenth century. The 50 provinces of the Spanish state are still trying to find a place in the evolving political system. The growth and ascendancy of regional government is leading to a simultaneous decline of provincial government.

The semi-federal organization of the Spanish state is reinforced by the role that representatives of regional government play in the upper house of the Spanish Cortes, the Senate. The present Partido Popular government is strongly dependent on regional support, which has led to further concessions to regionalization. In time this chamber may be transformed into a strong, territorial chamber able to play an influential role in matters concerning regional, provincial and local issues (*El Pais*, 3 October 1994, p. 12).

Regionalism in Spain: a multi-level party system

Josép M. Vallés coined the expression '*Espanas electorales*' (Electoral Spains) to characterize the coexistence of national and subnational electoral arenas (Vallés, 1989, pp. 33–5). At the moment Spain has a national electoral procedure for both houses of Parliament. This is complemented by the 17 regional electoral arenas with different party systems, and even different election dates. Until the late 1980s the levels of electoral geography did not interact. This changed in 1993, when the PSOE was not able to gain a clear majority in the Congress of Deputies and had to ask for support from the fourth largest party, the Catalan Convergence and Union (*Convergencia i Unio* – CiU). After the general elections of 3 March 1996, Convergence and Union adopted a strategy of supporting the largest party, the People's Party (*Partido Popular* – PP), which also enjoyed support in the Congress of Deputies from smaller regional parties of Aragon and Valencia (Amodia, 1996; Balfour, 1996, pp. 281–4).

The growing importance of regional electoral arenas accounts for the level of support for the two large national parties at regional level. In the Basque Country the dominant party is the National Basque Party (*Partido Nacional Vasco* – PNV), founded in the late nineteenth century as a product of Basque Nationalism. It represented a radical stand by demanding independence for the Basque country, a position which it continues to maintain. Nevertheless its actual strength was always dependent on the support of the regional branch of the Socialist Party. A coalition between these two main parties has existed since the early 1980s. The first regional elections in the Basque Country were held in the 1980s, leading to a victory for the PNV.

In political terms Catalan nationalism was more moderate and based its strength on the economic dominance of its bourgeoisie in Spain. The rebirth of Catalanism after the transition led to the subsequent dominance of the bourgeois-nationalist CiU. The CiU is a merger of two parties and recently divergences between the leader of *Convergencia Democratica de Catalunya*, Jordi Pujol, and the leader of *Unio Democratica de Catalunya*, Antonio Duran Lleida, re-emerged after the tenth party conference in December 1996. CiU lost the absolute majority in the last Catalan elections in November 1995. The second-largest party is the *Partit dels Socialistes de Catalunya* (PSC-PSOE) with 34 seats. Since the last elections the People's Party was able to become third-largest party by increasing its vote by 7 per cent to 13.09 per cent, gaining 17 seats.

These two autonomous communities differ considerably from all the other 14, where regionalist–nationalist parties are rather weak. In the other areas, party systems are dominated by the national parties, the PSOE and the PP. Until 1995 the strongest national party in regional elections was the PSOE. This changed in the regional elections of 28 May 1995 when seven out of ten former Socialist autonomous communities voted for a PP regional government. At the time of writing, eleven regional governments are dominated by PP, three by PSOE (Extremadura, Castilla-La Mancha, Andalucia) and a further three by regionalist–nationalist parties (*ABC*, 29 May 1995, pp. 61–96).

Regionalism has transformed Spain in the past two decades. It has led to changes of attitude amongst conservatives who previously regarded it as a threat to Spanish unity. A semi-federalist state is emerging which may lead to even more radical changes in the Spanish polity. On the whole, regional elections increased in importance as an indicator for the acceptance of government. Left and right were highly polarized at the national level, but not at the regional. So alternance between left and right was less problematic at the regional than at the national level. Indeed, polarization was beginning to disappear altogether in the 1990s. In this instance the regional area has proved to be an important antechamber for new political alternatives.

The recent debate on the autonomies

The recent debate on the autonomies in the Senate, on 10–12 March 1997, revealed that the demand for more autonomy by the 17 autonomous communities was being met by the government. In 1994 the Senate was formally transformed into a territorial chamber and a committee was set up to monitor the process of regionalization. Several reforms were implemented. A more generous system of financing for autonomous communities enshrining the co-responsibility of the autonomous governments was set up. The reform of the peripheral administration of the state LOFAGE (*Ley de Organisacion y Funcionamiento dela Administracion General del Estado*) was initiated. The participation of the autonomous communities in the European institutions was assured. Further development of the statutes is on the agenda. In the ten months of PP government over 80 devolution issues were dealt with, involving expenditure of 50,000 million pesetas. (*El Pais*, 11 March 1997, p. 16; *El Pais*, 13 March 1997, p. 15).

The historical regions are still demanding increased autonomy. Jordi Pujol's CiU has asked for parity between the capital in Castille and Catalonia. Shared sovereignty would entail a confederal system in language, culture and the civil code, regarded as the best way to preserve a distinct national identity. Parity would include a similar financial model to that of the Basque Country, according full sovereignty over taxation raised in the autonomous community.

Likewise, the president of the Basque government has pressed the coalition government to grant a more fully elaborated Economic Concertation Agreement to the Basque Country. The Basque nationalists, dominated by the PNV, were however against a debate on the statute of the Basque Country within a general discussion concerning all 17 autonomous communities.

The impact of structural funds in Spain

Spain has clearly profited from the Delors II package, with a tripling of structural funds, rising from ecu 15,087 billion to ecu 42,400 billion from the period 1989–93 to the period 1994–99 (European Commission, 1997, p. 45). Most of the funds were allocated to the modernization of public infrastructures, investment in human resources and the productive environment. The structural funds also reinforce the linkage between the regions and the European Union. They encourage the further devolution of competences to the region, help to improve and foster regional and local administration, and extend the time horizon of policy-making. The first and second Common Support Framework included the regional institutions in shaping the decision-making and implementation process of the European Union. The structural funds also encourage the inclusion of regional civil society in this process, through the permanent consultation of social partners represented in the Economic and Social Committees. They reinforce the decentralization and democratization of decision-making and the linkage between the regions and the European Union.

The regionalization debate and national populism in Portugal

Historically and politically, Portugal's experience is quite distinct from Spain's. Yet here too regionalization and the EU's regionalized territorial policies have altered and extended the rationale of governance.

The regionalization debate

The Portuguese constitution provides for the establishment of administrative regions (*regiõs administrativas*) with a regional government and regional assemblies, in part directly elected. The main task of the administrative regions is to co-ordinate efforts of local authorities and to intermediate between local and central government. At the moment the governing Socialist Party is trying to agree a compromise with the main opposition party regarding administrative regions. The regionalization debate is motivated by the idea that socioeconomic development between municipalities could be better coordinated and that decentralization would introduce more democratic accountability through elections (Oliveira, 1996, pp. 504–9). The main difficulty involves the legacy of the unitary state. Portugal has only ever had two tiers of governance; central and local. During the electoral campaign of 1995 the Socialist Party argued that regionalization would lead to a deconcentration and decentralization of the administration and make it more citizen-friendly; however the constitutional provision was never put into practice.

A 'framework law on the administrative regions' exists, approved by the Assembly of the Republic, Law no. 56/91 of 13 August. The main difficulty is the efficient division of the national territory. There are proposals to divide it into five macro-regions or nine smaller ones (Oliveira, 1996, pp. 500–1). For the moment the administration of the territory is coordinated by (devolved Regional Coordinating Commissions (*Comissoes Coordenadoras Regionais* – CCR) and the intermediary decentralized tier is the pre-revolutionary administrative unit of the district (*distrito*). The provincial assembly is filled by the local authorities through co-option, and presided over by a civil governor (*governador civil*) appointed by the government (Magone, 1997, p. 63). The fear that decentralization will lead to the disintegration of the traditionally unitary Portuguese state was the main reason why former Prime Minister Cavaco Silva opposed the regionalization process (*O Publico*, 30 July 1994, pp. 2–5). Nevertheless, the population in the north is acquiring a regional consciousness which may lead to stronger demands for the establishment of proper administrative regions. The mayor of Oporto established international links across the border with Manuel Fraga Iribarne, president of the Xunta de Galicia. Trans-border cooperation such as that between Alentejo and the Spanish autonomous community of Extremadura is supported by European programmes such as INTERREG. With increasing European integration such strategies of regional development have become increasingly common (Cruz, 1995, pp. 484–5; Covas, 1995, pp. 72–8).

The process of decentralization is still in its early stages, and the central administration still makes up the largest proportion of the civil service. In 1935 all 25,588 civil servants worked for the central administration. In 1968 the ratio was 78.9 per cent central to local. By 1979 the number of civil servants had almost doubled to 372,295 – with a ratio of 84.3 per cent central to local. In a survey conducted in 1988 by the Secretariat of Administrative Modernization (*Secretariado da Modernizacao Administrativa* – SMA), the total number of employed persons emerged as 513,770 with a ratio of 46.3 per cent in the central administration, 32.44 per cent at district level and 21.44 per cent in local administration. By 1991, however, this trend had reversed: the 624,000 civil servants were divided into 87.18 per cent central and 12.82 per cent local. In this sense the democratization process has led to a deconcentration of services accompanied by signs of centralization of administration, though that may be reversed in the future. What is at stake is not so much deconcentration, but decentralization. The decentralization of the political system would enhance the possibilities of citizens to influence policy-making.

The long-term strategy of regional plans of development is intertwined with the European integration process. For the moment the lack of regional authorities makes the Portuguese central government the central 'gatekeeper' between the European Union and the regions. The CCRs are central governmental bodies lacking the democratic legitimacy of administrative regions (Syrett, 1994a, 1994b, p. 54 and pp. 59–60). Regionalization is predominantly linked to economic and social development. Nevertheless all these different regions are culturally distinct, even if they share the same language and historical legacy.

The emergence of national populism

Regionalization is also a means to re-create democracy in spatial terms. By designing more citizen-friendly integrative institutions it is hoped that evenness of development across the territory will be achieved. Recent developments indicate that there will be a referendum on the nature of the administrative regions.

After the collapse of the authoritarian regime no radical right-wing parties were able to establish themselves. For years the Party of Christian Democracy (*Partido de Democracia Crista* – PDC) contested elections without significant success. Conversely, radical organizations such as Kaulza Arriaga's short-lived Independent Movement of National Reconstruction (*Movimento Independente de Reconstrucao Nacional* – MIRN) in the early 1980s or the violent Movement of National Action (*Movimento de Accao Nacional* – MAN) in the late 1980s were unable to mobilize the population for nationalistic or xenophobic purposes. The moderation of the Portuguese electorate has prevented the emergence of radical groups on the right and left (Pinto, 1995, pp. 108–19).

If there is a right-wing, national populist party it is the People's Party, formerly the Christian Democratic Social Centre (*Centro Democratico Social* – CDS). Its nationalist tone is directed against the European Union. The Portuguese Partido Popular (PP) transformed itself into an anti-Maastricht Treaty party, opposing further integration on the grounds of loss of sovereignty. Since 1992 the PP has demanded a referendum for all European treaties that will affect Portugal's future. In the elections for the European Parliament on 12 June 1994 the PP adopted national sovereignty as its main issue, campaigning for the preservation of independence from supranational governance and the maintenance of national parliament as the prime decision-maker. In spite of this, the PP supports the flow of structural funds from the European Union to Portugal as compensating policies for the opening-up of the Portuguese market (Centro Democratico e Social-Partido Popular, 1994). Between 1991 and 1995 the party was able to more than double its electoral representation. In terms of electoral strategy the party has mobilized those who lose out to European integration in the urban and rural areas, using a populist nationalist language.

During a party congress in the northern town of Povoa do Varzim on 23–24 January 1993, Manuel Monteiro accomplished a change of party strategy, convincing 700 PP delegates to vote in favour of a more social market economy and against regionalization (*Expresso*, 23 January 1993, A3; *Expresso*, 30 January 1993). The CDS-PP was excluded from the European People's Party in 1992 on the grounds that it did not follow the same European policies of further integration as the other parties. Nevertheless the party was able to join the Alliance of European Democrats (*Expresso*, 3 April 1993, A3). After their poor local elections results on 14 December 1997 the new leader, Paolo Portas, remained opposed to the regionalization process. Civic associations, such as the 'Movement for a United Portugal' (*Movimento Portugal Unico*), are also campaigning against regionalization. And a new extreme right-wing party, the Nationalist Portuguese Party (*Partido Nacionalista Portugues* – PNP), with a programme based on traditional values of fatherland, morality and authority, is also an opponent of the regionalization process (*Expresso*, 23 May 1998).

Yet, on the whole, PP is a reflection of moderate anti-establishment feelings. Analysed at a deeper level it voices the problems encountered by some sectors of the population in coping with transformed territorialities. The rationale of European integration is causing disruption in the pre-existing social fabric, producing support for this highly conservative party. In a sense, however, assuming that Portuguese democracy proves strong enough to integrate such marginal positions and concerns into the political system, the emergence of PP can only be regarded as a positive element in the construction of Portuguese politics.

The impact of structural funds in Portugal

Between 1986 and today, the impact of the structural funds in Portugal has been the single most important factor in pushing regional development. Medium-term programming rendered public-policy thinking more amenable to integration and required a growing mobilization of the population in favour of it. The whole territory of Portugal was eligible for the allocation of structural funds. The funds allocated for the period 1989–93 were about ecu 9.43 billion; for the period 1993–9 they increased to ecu 17.64 billion. Funds were invested in infrastructure, human resources, the productive environment, education (through the Specific Programme for the Development of the Portuguese Education System – *Programa Especifico para o Desenvolvimento do Sistema Portugues de Educacao* – PRODEP – and the Specific Programme for the Development of the Portuguese Industry – *Programa Especifico para o Desenvolvimento da Industria Portuguesa* – PRODIP). The structural funds were a major factor in improving the quality of the administration and the processes of monitoring and evaluating interventions in conjunction with the project manager (European Commission, 1997, p. 117). Thus, by mobilizing civil society in the periphery, the structural funds have enhanced democratic structures in the regions.

Conclusion: the end of marginality?

The reconstruction of the Euro-polity has coincided with a reconstruction of the Spanish and Portuguese polities. For the Iberian countries European integration has entailed adjustment to a different kind of administrative culture, based on compromise (Abeles and Bellier, 1996). This is permeating the structures of the two new democracies. Patrimonial and clientelistic forms of interest intermediation and territorial representation are being replaced by accountable administrative and political forms of intermediation and representation. Without the regionalization of Iberian politics, echoing and enhancing that of the European Union, this democratization could not have taken place.

Yet, at the same time as these developments in Spain and Portugal, national identities at supranational and subnational levels are being questioned elsewhere in Europe. The emerging pattern is one of multiple identities which may overlap but do not completely match. The territorial policies of the European Union attempt to achieve some convergence of the political structures of the fifteen countries while maintaining the distinctiveness of individual countries. Yet this cannot be done without the people of the fifteen countries having a voice for their concerns and fears. The need to overcome the democratic deficit at the European Union is primarily, therefore, a need to mobilize the population for a discussion about their identities. It cannot be met without an injection of democracy articulated in spatial terms.

Until the mid-1970s Spain and Portugal were considered marginal to Europe. Since then, Europeanization and democratization have transformed both polities into models for a Europe of democratic transition and consolidation. What has been happening is a transition from formal to qualitative democracy characterized in part by spatial democratization of the political system. Spatial democratization of the political system is one element in the qualitative enhancement of democracy. Furthermore, in a contemporary setting which has been dubbed the era of 'postmodernity', when identities and rationalities, political and otherwise, are bound for perpetual questioning, the process of shifting and reorganizing boundaries in Iberia may prove to be an important exercise for the eventual cultural integration of the European Union.

References

Abeles, Mark and Bellier, Irene (1996) 'La Commission Européenne du Compromis Culturel à la Culture Politique', *Revue Française de Science Politique*, **46**(3), 431–56.

Amodia, José (1996) 'Spain at the Polls: the General Elections of 3 March 1996', *West European Politics*, **19**(4), 813–19.

Balfour, Sebastian (1996) 'Bitter Victory, Sweet Defeat: the March Elections in Spain', *Government and Opposition*, **31**(3), 275–87.

Barbera, Rita (1996) 'Los Municipios Espanoles', *El Pais: anuario 1996*, p. 120.

Belloni, Frank P. (1994) *The Single Market and Socio-Economic Cohesion in the EC: Implications for the Southern and Western Peripheries* (Bristol: University of Bristol, CMS–Occasional Paper, no. 8), February.

Bonime-Blanc, Andrea (1987) *Spain's Transition to Democracy* (Boulder, CO: Westview).

Brassloff, Audrey (1991) 'Spain: Democracy and Decentralisation', in A. M. Brassloff and W. Brassloff (eds), *European Insights. Postwar Politics, Society and Culture* (Amsterdam: Elsevier), pp. 57–68.

Centro Democratico e Social-Partido Popular (1994) *Viva Portugal* (Lisbon: Manifesto Eleitoral).

Centro de Informacao Cientifica e Técnica (1987) *Administracao. Factos e Numeros* (Lisbon: Centro de Informacao Cientifica e Técnica).

Commission of the European Communities (1992) *Towards Sustainable Development. A Programme of the European Community in Matters of Environmental Policy and Action and Sustainable Development.* COM(92)-23 final, 2 vols (Brussels), 26 May.

Covas, Antonio (1995) 'La cooperation transfrontalière entre regions sous-developpées: le cas d'Alentejo et d'Extremadure (Espagne)', *Pole Sud*, **3** (Autumn), 72–8.

Cruz, Manuel Braga da (1995) *Instituicoes Politicas e Processos Sociais* (Venda Nova: Bertrand).

Engel, Christian (1993) 'Die regionen im Netzwerk Europaeischer Politik', in U. Bullman (ed.), *Die Politik der Dritten Ebene* (Baden-Baden: Nomos), pp. 91–109.

European Commission (1994) *Guide to the Community Initiatives 1994–99* (Luxemburg: Office for Official Publications of the European Community).

European Commission (1995) *Cohesion and the Development Challenge Facing the Lagging Regions* (Luxemburg: Office for Official Publications of the European Communities). Com(95) 111 of 29 March.

European Commission (1997) *The Impact of Structural Policies on Economic and Social Cohesion in the Union, 1989–99. A first assessment presented by country* (October 1996) (Luxemburg: Office for Official Publications of the European Communities).

Ferreira, José Medeiros (1983) *Ensaio Historico sobre a Revolucao de 25 de Abril* (Lisbon: Casa da Moeda).

Figueiredo, Antonio de (1975) *Cinquenta Anos de Ditadura em Portugal* (Lisbon: Dom Quixote).

Gillespie, Richard (1995) 'Perspectives on the Reshaping of External Relations', in R. Gillespie, F. Rodrigo and J. Story (eds), *Democratic Spain: Reshaping External Relations in a Changing World* (London: Routledge), pp. 196–201.

Goybet, Catherine (1997) 'Agenda 2000: La Commission Européenne prépare L'Europe Élargie à l'Est', *Revue du Marché Commun et de l'Union Européenne*, **411** (September–October), 509–11.

Graham, Robert (1984) *Spain: Change of a Nation* (London: Michael Joseph).

Heywood, Paul (1995) *The Government and Politics of Spain* (Basingstoke: Macmillan).

Hooghe, Liesbet and Marks, Gary (1996) 'Europe with the Regions: Chan-

nels of Regional Representation in the European Union', *Publius*, **26**(1), 73–91.
Hrbek, Rudolf (1989) 'Nationalstaat und Europaeische Integration. Die Bedeutung der nationalen Komponente fuer den Integrationsprozess', in P. Haungs (ed.), *Europaeisierung Europas?* (Baden-Baden: Nomos), pp. 81–108.
Jochimsen, Reimut (1998) 'Europa 2000 – Herausforderungen fuer die EU nach dem Vertrag von Amsterdam', *Integration*, 21 (1/98), 1–11.
Kalbfleischer-Kottsieper, Ulla (1993) 'Der Auschuss der Regionen-ein neuer Arbeitsperspektiven', in U. Bullmann (ed.), *Die Politik der Dritten Ebene* (Cologne: Nomos) pp. 134–143.
Linz, Juan (1978) 'Early State-Building and Late Peripheral Nationalisms Against the State: the Case of Spain', in S. N. Eisenstadt and R. Stein (eds), *Building States and Nations*, vol. II (London: Sage), pp. 31–116.
Loughlin, John (1996) 'Representing Regions in Europe: the Committee of the Regions', *Journal of Regional and Federal Studies*, **6**(2), 147–65.
Magone, José M. (1996) *The Changing Architecture of Iberian Politics. An Investigation on the Democratic Structuring of Political Systemic Culture in Semiperipheral Southern European Societies* (New York: Edwin Mellen Press).
Magone, José M. (1997) *European Portugal. The Difficult Road to Sustainable Democracy* (Basingstoke: Macmillan; New York: St Martin's Press).
Maravall, José Maria (1982) *Transition to Democracy in Spain* (London: Croom Helm).
Marks, Gary (1993) 'Structural Policy and Multilevel Governance in the European Community', in A. W. Cafruny and G. C. Rosenthal (eds), *The State of the Community*, vol. 2: *The Maastricht Debates and Beyond* (Boulder, CO.: Lynne Rienner), pp. 391–410.
Marks, Gary (1996) 'An Actor-Centred Approach to Multi-Level Governance', *Journal of Regional and Federal Studies*, **6**(2), 20–38.
Oliveira, César (1996) 'A questao da regionalizacao', in C. Oliveira (ed.), *Historia dos Municipios e do Poder Local* (Lisbon: Temas e Debates), pp. 495–509.
Os Portugueses e a Politica (1973) Estudos IPOPE 2 (Lisboa: Moraes Editores).
Pina, Antonio Lopez and Aranguren, Eduardo L. (1976) *La Cultura Politica de Espana de Franco* (Madrid: Taurus Ediciones).
Pinto, Antonio Costa (1995) 'The Radical Right in Contemporary Portugal', in L. Cheles, R. Ferguson and M. Vaughan (eds), *The Far Right in Western and Eastern Europe* (London: Longman), pp. 108–28.
Schmitter, Philippe C. (1996a) 'Imagining the Future of the Euro-Polity with the Help of New Concepts', in G. Marks, F. W. Scharpf, P. C. Schmitter and W. Streeck (eds), *Governance in the European Union* (London: Sage), pp. 121–50.
Schmitter, Philippe C. (1996b) 'How to Democratize the Emerging Euro-Polity: Citizenship, Representation, Decision-Making'. Mimeographed manuscript. Stanford University and Instituto Juan March, revised version, April.
Secretariado para a Modernizacao Administrativa (1989) 'Réponse Portugaise au Questionnaire sur les fonctions publiques Européennes', *Revue Française d'Administration Publique*, no. 55, July–September (mimeographed version).

Solé-Villanova, Joaquim (1989) 'Spain: Developments in Regional and Local Government', in R. Bennett (ed.), *Territory and Administration in Europe* (London: Pinter).
Syrett, Stephen (1994a) *Local Development. Restructuring Locality and Economic Initiative* (Aldershot: Avebury).
Syrett, Stephen (1994b) 'Local Power and Economic Policy: Local Authority Economic Initiatives in Portugal', *Regional Studies*, **28**(1), 53–67.
Tranholm-Mikkelsen, Jeppe (1991) 'Neo-functionalism: Obstinate or Obsolete? A Reappraisal in the Light of the New Dynamism of the EC', *Millennium: Journal for International Studies*, **20**(1), 1–22.
Vallés, Josép M. (1989) 'Entre la Regularidad y la Indeterminacion:Balance sobre el Comportamiento Electoral en Espana (1977–1989)', in J. V. Beneyto, (ed.), *Espana, a debate. I: La Politica* (Madrid: Editorial Tecnos), pp. 27–43.
Wallace, Helen (1996) 'Politics and Policy in the EU: the Challenge of Governance', in Helen Wallace and William Wallace (eds), *Policy-Making in the European Union* (Oxford: Oxford University Press), pp. 3–36.

9
Euroscepticism in the Ideology of the British Right

Christopher G. Flood

Introduction

There is a rapidly growing literature on the politics of Euroscepticism, particularly in its right-wing expressions, but less attention has been paid to its structure as an ideology. Euroscepticism constitutes an ideology in the neutral meaning of the term, as exemplified by Martin Seliger's short definition of ideologies as 'sets of ideas by which men [and women] posit, explain and justify ends and means of organized social action, and specifically political action, irrespective of whether such action aims to preserve, amend, uproot or rebuild a given social order' (1976, p. 14). However, Euroscepticism needs to be defined as a partial ideology in the sense that it does not offer a comprehensive, potentially universalizable view of man and society in the way that, for example, liberalism and socialism do. It focuses on a particular dimension of a particular society, or set of societies, which in this case is the EU.[1] Like all ideologies it does not have hard, impermeable boundaries. It can be combined with other ideologies. That is why there are left- and right-wing versions of it in many member states of the EU. Prior to the election of 1997, the variants of it articulated by large sections of the Conservative Party attracted particular attention because they involved a governing party in disarray, but we know there are sections of the Labour Party or left-wing groupings which are Eurosceptical, just as Euro-enthusiasm can be found in varying versions right across the political spectrum.

This chapter examines the ideological aspects of right-wing Euroscepticism in Britain. It does not enter into the details of technical arguments against particular aspects of EU integration. Rather, its aim is to map the major contours of the site. Equally, it does not seek to evaluate the claims put forward by Eurosceptics. The underlying premise is that, while it is possible to make more or less objective assessments of particular features of the EU and its operations, the complexities of the processes involved in European integration, as well as the imponderables involved in hypothesizing alternatives to integration or even of predicting the future course of integration itself, mean that the overall costs and benefits of the process are not calculable by any objective measurement. Political debate on the issue is therefore coloured particularly heavily by ideology.

My discussion takes account of the functions, and the increasing importance, of Euroscepticism as a component of the different ideological currents of the right. It sheds light on the contribution of Euroscepticism to revitalizing right-wing nationalism in response to perceived threats of absorption (control drained off to supranational bodies) and/or swamping (uncontrollable influxes of persons, goods, cultural products, etc.) which would reduce Britain to a marginal role in Europe and the world. The instrumental function of Euroscepticism in inter- and intra-party competition is also taken into consideration.

Although I make brief reference to developments prior to the 1990s, the central focus is on Eurosceptical positions adopted during and since the period of intense debate over ratification of the Treaty of European Union (TEU) negotiated at Maastricht in 1991. The stridency and weight of recent ideological reactions against European integration are products of a particular context. They owe a great deal to the long-term effects of the oil shocks and economic stagnation of the 1970s which did much to discredit the post-war social-democratic, Keynesian consensus, thus preparing the way for the rise of neo-conservatism. But they have been heightened by the seismic international effects of political and economic change during the 1990s, with the collapse of the post-war bipolar block system and the accelerated progress of transnational capitalism, which have raised new possibilities but also anxieties even among those who are ideologically most in harmony with much of this development. At the same time, the promotion of a new, dynamic phase of European integration from the mid-1980s to the mid-1990s by Jacques Delors, François Mitterrand and Helmut Kohl accentuated the supranational, technocratic dimension of the organization. But for many right-wing politicians in Britain it was the TEU far more than the Single European Act (SEA) of 1986 that crystallized the perception of an imminent threat to be countered by vigorous campaigning before it was too late.

The primary sources for the chapter are the writings and speeches of politicians and other elites, rather than grassroots party members or voters. In order to examine different degrees and kinds of right-wing Euroscepticism, the analysis is based on material produced by representatives of the Conservative Party, the UK Independence Party, the Referendum Party (now fused with the Democracy Movement), the crypto-fascist British National Party, the post-fascist New Democrats and Third Way, plus various pressure groups. With the exception of the Conservative Party these are all fairly small or very small organizations, but they illustrate the range of right-wing Euroosceptical positions – in any case, the size of an organization has no necessary correlation with its ideological substance.

The context: Euroscepticism in British right-wing politics during the 1990s

In Britain right-wing hostility to European integration was not a new phenomenon in 1992, nor was its capacity to divide the Conservative Party. Admittedly, overt opposition within the parliamentary Conservative Party at the time of Macmillan's attempts to negotiate Britain's entry to the EEC in 1961–63 had been relatively contained. This had remained the case at the time of Wilson's attempt in 1967 and of Heath's successful negotiation of membership in 1971–73 and of the 1975 referendum on continued membership (Beloff, 1996; Kitzinger, 1973; Morris, 1996). Nevertheless, it had not implied wholehearted commitment to European integration among the mainstream of the party – far from it. If that had been the case, Macmillan and Heath would not have felt it expedient to play down the long-term constitutional implications of membership while emphasizing the putative economic advantages. It is true, of course, that from the mid-1970s to the mid-1980s the party had gained the reputation of being the 'party of Europe' in contrast to the Euroscepticism which dominated the Labour Party at that time. But although the party contained a minority of Europhile supranationalists, just as it also contained a minority of Europhobic opponents of membership, the commitment among other Conservatives tended to be lukewarm, protective of national sovereignty, and supportive of further integration in the economic sphere alone. The latent divisions within the party emerged more clearly in the late 1980s as the process of integration quickened in the wake of the SEA of 1986. Among the major factors which contributed to Margaret Thatcher's removal from the leadership of the party, and hence from the prime-ministership, in 1990 was the belief held by a number of senior ministers and backbench MPs that her increasingly vocal antagonism towards further integration after the passage of the SEA – notoriously heralded by her speech to the College of Europe in Bruges on 20 September 1988 – was becoming a serious liability to Britain's interests in Europe. However, Lord Beloff (1996, p. 92) may well be right when he suggests that Michael Heseltine's failure to gain the succession indicated the limits of support for Eurofederalism within the party.

By the mid-1990s the processes set in train by the TEU, building on the major advances already stemming from the SEA, had given particular intensity to the dispute within the Conservative Party. The antagonism was also spurred by the forced departure of sterling from the Exchange Rate Mechanism of the European Monetary System on Black Wednesday, 16 September 1992. It left Eurosceptics blaming Germany for its refusal to lower interest rates and claiming this as a timely warning of the dangers of the whole European monetary project. More generally, the problem was aggravated by weak leadership and an apparently vacillating stance on Europe under John Major, whose parliamentary majority was small enough to increase the leverage of the twenty-plus dissident MPs, notwithstanding Major's success in securing ratification of the TEU after bitter intra-party conflict right up to cabinet level in 1993. The party's weak showing of 28 per cent of votes cast in the 1994 EP election (a loss of 1.1 million votes compared with its already poor performance in 1989) added to the air of decay, even if the motivation for voters' choices was heavily influenced by domestic rather than European issues (George, 1994, pp. 212–16).

Major's effort to strike an assertive stance by vetoing the appointment of the Eurofederalist Jean-Luc Dehaene as President of the Commission in 1994 was undermined by the fact that he did not veto Jacques Santer, despite the fact that Santer's own integrationist credentials were not significantly weaker than Dehaene's. With the government bowing to Eurosceptics in terms of its rhetorical posture – for example, on fishing quotas or regarding the ban on British beef exports following the 1996 crisis over BSE and its possible transmission to humans – yet still refusing to rule out British participation in EMU, the travails of the party over Europe continued up to the time of the 1997 election. Furthermore, the balance of support among Conservative parliamentarians appeared to be shifting increasingly in favour of Eurosceptical positions. The politics of right-wing Euroscepticism often dominated the headlines in the press and provided material for endless discussion in the broadcast media. Contrary to its traditional pride in maintaining an appearance of loyalty and discipline, the Conservative Party exhibited its internal divisions in an increasingly overt way. Sceptical views were also promoted strongly in the Conservative-supporting national press, ranging from *The Times* and *Daily Telegraph* (with their Sunday counterparts) to the *Daily Mail* and *Daily Express* (also with Sunday counterparts) to the *Sun* (with its Sunday counterpart, the *News of the World*), not to mention the weekly *Spectator*. They also spawned a substantial number of books, often with revealing titles, such as *Europe: The Crunch* (Cash, 1992), *A Treaty Too Far* (Spicer, 1992), *The Castle of Lies* (Booker and North, 1996), *The Tainted Source* (Laughland, 1997), *Treason at Maastricht* (Atkinson and McWhirter, 1995), *The Rotten Heart of Europe* (Connolly, 1996) or *Britain Held Hostage* (Jenkins, 1997). There were even futuristic thrillers, set in the corrupt European superstate of the mid-twenty-first century (Roberts, 1995; Palmer, 1997).

The categorical defeat of the Conservative Party (31 per cent) by a cautiously Euro-enthusiastic Labour Party (44 per cent) in the general election on 1 May 1997 implied that Conservative claims to stand for firmness against European federalism did not hold a decisive appeal in comparison with other factors – though Eurosceptics would argue that the party did not take a clear enough stance on the issue (for example, Holmes, 1998). It was widely supposed that the ill-concealed divisions within the parliamentary Conservative Party over Europe helped to reinforce a public perception that it would be incapable of coherent government in the future. In the weeks following the defeat, the parliamentary party, now reduced in number from 343 MPs to 157, showed that a majority of its members remained firmly committed to a tough Eurosceptical line for the future. Of the six candidates for the leadership, four were associated with hardline Euroscepticism (John Redwood, Michael Howard, Peter Lilley, William Hague) and another (Stephen Dorrell) had trimmed in that direction over the previous months. Only Kenneth Clarke, former Chancellor of the Exchequer, was more or less avowedly Europhile. Clarke, though shown by opinion polls to be the candidate most favoured by a majority of party members and by the wider public, was soundly defeated in the third round by William Hague, who had proclaimed his Eurosceptical credentials unequivocally on the symbolic issue of European Monetary Union (EMU) by ruling out British entry for the lifetime of two parliaments. This demonstrated that the parliamentary party had not taken the recent electoral failure as a signal for it to change tack on European integration – quite the reverse. The tone was set in the wake of the Amsterdam European summit in June 1997 when Hague attacked Tony Blair, the Prime Minister, for giving away British sovereignty in important areas during the final negotiations and demanded in populist fashion that Blair should submit the treaty to a referendum (Hague, 1997). The Eurosceptical stance has been sustained by the party's leadership since that time (for example, in Hague, 1998; Portillo, 1998).

The issue of Europe had meanwhile provided a vehicle for more radical, nationalist elements on the edges of, or to the right of, the Conservative Party. These included the Referendum Party founded and massively funded by the Anglo-French businessman Sir James Goldsmith, leader of the Europe of Nations group in the European Parliament; the United Kingdom Independence Party, founded by Alan Sked; and new or old pressure groups, such as the Campaign for an Independent Britain, the Critical Europe Group, the Freedom Association and the Anti-Common Market League. A virulent Europhobia was promoted by the supposedly apolitical heritage magazine *This England* under the banner: 'Don't Let Europe Rule Britannia'. On the extreme right the issue was bread and butter for the National Democrats, Third Way and the British National Party, all of them descended from the National Front. The very weak support for single-issue Europhobic parties and for the extreme right in the election brought a pause in the campaign against further European integration. But it did not, of course, indicate a change of ideological perspective on the part of those groups.

Eurosceptical positions

There are different degrees of right-wing Euroscepticism in Britain, as well as variations of emphasis. In terms of policies, three broad tendencies can be distinguished within the British right. The least radical of the three was encapsulated in the official line taken by the Conservative government from 1992 to 1997, despite the increasingly negative emphasis adopted by many ministers, including the Prime Minister, John Major, in the run-up to the 1997 election. The positions set out in the white paper in March 1996 for the IGC reflected this thinking (FCO, 1996; Rifkind, 1996a, 1996b). The stance was essentially defensive. It involved an assumption that the Maastricht Treaty took the integration of the EU as far as it ought to go – too far, in some areas. Britain should concede additional decision-making powers to the Community only in cases of demonstrable national benefit. While the white paper endorsed the aspiration to closer cooperation and friendship, it explicitly rejected 'an ever closer Political Union in the sense of an inexorable drift of power towards supra-national institutions, the erosion of national parliaments, and the gradual development of a United States of Europe' (FCO, 1996: §5). This position meant, for example, preserving Britain's opt-out on the Social Chapter. It meant resisting new powers for the Commission or the EP and resisting the extension of QMV into new areas in the Council of Ministers. It involved resisting the incorporation of the whole or parts of the existing Third Pillar (Justice and Home Affairs) under EU jurisdiction. Although it allowed some modest proposals for promoting the development of the CFSP – including the appointment of an official to represent the EU's foreign policy to the outside world – it entailed rejecting a more integrated approach, defending the existing intergovernmental processes and explicitly rejecting the idea that the Western European Union (WEU) should become a vehicle for integrated EU defence. It meant keeping options open on EMU, while defending the principles of subsidiarity, decentralization, deregulation, variable geometry and intergovernmentalism against supranationalism wherever possible. The aim was to put Britain's national interest at the forefront of all considerations and to support a wider but not a deeper EU.

A more radical Eurosceptical position than that of the former government argues that Maastricht was a treaty too far. While the provisions relating to the SEM should be retained, there should be a cap on Britain's budgetary contributions and those elements which eroded national sovereignty in other spheres should be rolled back. So, for example, the power of the ECJ should be curbed, as should the powers of the Commission. More generally, the weight of EU regulation should be massively reduced. Beyond the maintenance of the single market, all other areas of shared activity should be pursued on an intergovernmental basis. This more radical approach, involving demands for revision of the TEU, is overtly espoused by many right-wing Conservative MPs and ex-MPs, such as Bill Cash, John Redwood, Michael Howard, John Biffen, Lord Tebbit, Edward Leigh and Sir Michael Spicer. It is heavily promoted by the Thatcherite Conservative Way Forward Group, among others. It has also been gaining ground across the centre of the party.

At the Europhobic end, there are those who argue that Britain should withdraw from the EU altogether. They maintain that the present balance of political power within the EU makes it impossible for Britain to get what it wants and that there are no other conclusive reasons for remaining a member, whereas there are a number of increasingly important gains to be derived from non-membership, albeit, perhaps, with an association agreement. The idea of withdrawal is rarely articulated in public by Conservatives, as yet, but is not repulsive to some right-wingers such as Norman Lamont, Sir George Gardiner, Tony Marlow and Teresa Gorman. It has been promoted by ginger groups, such as the Campaign for UK Conservatism and the Monday Club. Beyond the Conservative Party, the rejectionist position undoubtedly appealed to many supporters of the Referendum Party – though not to Sir James Goldsmith himself, it seems (Goldsmith, 1994), despite the increasingly feverish tone of his attacks on the EU during the months before the general election (for example, Goldsmith, 1996a, 1996b). Rejectionism is the defining position of the UKIP, as well as the Campaign for an Independent Britain, the Critical Europe Group, and the Anti-Common Market League. It is espoused by the Freedom Association and a number of more obscure pressure groups – not to mention the venomous campaign in *This England*. It has given rise to websites with evocative titles, such as Eurofighter, Freedom Fighters and Free Britain. On the extreme right it is essential fodder for the neo-fascist National Democrats, Third Way and the British National Party.

It goes without saying that these three tendencies are not hermetic blocks. The reality is that they shade into each other, and individuals may shift between them at different times or simply equivocate. For example, if convinced that there was no further hope of rolling back the integration process, a Maastricht-revisionist could become an EU-rejectionist. That attitude was suggested during the controversy over ratification of Maastricht, when a senior Conservative MP, Michael Spicer, argued that the British position in opposing political union ought to be: 'Above the line anything goes; below it we walk away' (1992, p. 193). Nevertheless, although there can be changes and exceptions, the shading across from moderate Euroscepticism to Europhobia tends to correlate roughly with the shading between moderate right and extreme right.

Ideological arguments and premises

Notwithstanding the differences of policy between different Eurosceptical groupings, many arguments against the current development of the EU are broadly shared by all groups. The most common of these can be summarised as follows:

1 The EU is on the way to being a centralized superstate, absorbing the national sovereignty of its member states. This is an outdated and unacceptable goal. The fate of the USSR and Yugoslavia shows what happens when diverse peoples are artificially bound together too tightly without adequate expression of their national identities and aspirations.

2 The EU is undemocratic, bureaucratic, inefficient and largely unaccountable to the peoples of its member states but democratizing it by increasing the powers of the European Parliament is not the answer since that involves further transfer of sovereignty to a remote body having little connection with the citizens.

3 The project for European Monetary Union and the European Central Bank will not only be economically ruinous but it will further consolidate the movement towards a centralized political system by effectively removing member states' control of their finances and of their national economies, which will then function under central authority.

4 It is unacceptable that European laws should have primacy over national laws. Moreover, the European Court of Justice has far too much power to reinterpret the treaties and other European legal instruments in a federalist direction.

5 The Schengen Agreement was a step too far in the relaxation of national border controls. Not only is it an affront to a vital sphere of national sovereignty, but it is also a recipe for laxity and failure to curb the problems of asylum, immigration and international crime. It is even more undesirable that matters of immigration and asylum should be transferred under the competence of the Commission.

6 To base a Common Foreign and Security Policy (CFSP) on Qualified Majority Voting (QMV) would be unacceptable and could only lead to division, given the vital national interests of member states involved in these fields. It would be equally undesirable

for the WEU to be subsumed as the integrated defence arm of the EU: NATO already provides an adequate structure for defence and security.

7 The obsession with deepening integration is necessarily at the expense of the urgent task of widening to embrace the nations of Central and Eastern Europe, since the conditions of entry become increasingly rigorous. The risk is that these states will dissolve into economic, ethnic and political chaos without the support that they need from the West.

What underlies these arguments? I have summarized them in their most abstract, general forms. Sometimes they are presented in this way in the discourse of Eurosceptical politicians and publicists. But in the cases of arguments (1)–(6), at least, they are more often stated in terms of Britain's national interests, while (7) reinforces the claims under (1)–(6). Britain's autonomy, Britain's identity, Britain's role in the world – these are the concerns which stir Eurosceptics of the right. The preoccupation is captured in the title of Norman Lamont's book, *Sovereign Britain* (1995), or in those of the other works cited earlier. The essence of the issues can be grasped in terms of a dialectic of marginality and centrality. On the one hand, the EU is seen as a structure within which the motive forces are disposed in such a way that the marginal geographical location of the UK in relation to the centre will necessarily be mirrored by the relegation of the UK to a marginal position in the EU's principal spheres of activity. Marginality in this sense means a two-fold process of erosion of well-being. As the EU continues its integration, it absorbs British sovereignty by drawing powers of decision and implementation into its central bodies. It demands massive financial contributions. It threatens to drain off what remains of Britain's economic and cultural autonomy. This, then, is a process of absorption. Its corollary is the process of swamping, which can occur in various ways. The central apparatus of control deluges the marginal territory with laws and regulations. In the arguments of Eurosceptics the principle of free movement of persons, coupled with lax approaches to immigration and asylum, opens up terrifying prospects of vast influxes of impoverished foreigners eager to batten onto our welfare system and our already saturated employment markets. For some extreme right-wingers, as we shall see later, even the principle of the free market in goods and services is an intolerable threat to Britain.

However, in the eyes of Eurosceptics Britain need not be

marginalized in this way. It may be on the edge of the EU in both a literal and a figurative sense, but the EU is only one among many sets of states in the international system. Britain can be understood as being drawn outwards towards the non-European world. This has at least two important aspects. In relation to the countries of its own former empire, Britain is not so much the edge of Europe, but the centre of the network of diplomatic, economic and cultural relationships. It can even be fondly imagined by some Eurosceptics that in a symbolic sense Britain remains the hub of the English-speaking world. Another way of looking at the position is to see Britain in its place as a privileged intermediary between Europe and the USA, now the undisputed power centre of the planet, which bestows a form of centrality by proxy on the go-between. Thirdly, Britain can capitalize on its outer/inner interface with continental Europe in order to attract inward investment from non-Europe so that the function of surrogate producer allows the UK to shift the axis of Europe in its favour. All of these positional and dispositional factors lead to the idea that Britain should retain an autonomous, semi-detached relationship with the EU.

How do these considerations intersect with other aspects of right-wing ideological thinking? The Conservatives can be taken as a starting-point. The party has always been a broad church. It is the heir to rich and varied ideological traditions which have given rise to at least three major tendencies within the party in the post-war period (subject, of course, to the usual caveats concerning such broad categorizations). These tendencies can be labelled progressivist, traditionalist and individualist (Whiteley *et al.*, 1994). On the left the progressivist, one-nation tendency, which considers itself heir to the Disraelian spirit of social reform, is closely associated with post-war acceptance of Keynesian management of the mixed economy and the welfare state. This tendency, which increasingly lost its dominance within the party from the early 1970s onwards, includes those who have given pragmatic, or even discreetly enthusiastic, support to British participation in European integration.

Of the other two tendencies, the traditionalist current is heir to the reactionary, nationalist strand of thought which looks back nostalgically to the days of empire and to the idea of Britain as a great power with global interests and responsibilities. This was the tendency on the hard, authoritarian right of the party which denounced decolonization as a sell-out, defended the white separatist regimes in Rhodesia and South Africa, objected to coloured immigration to Britain, defended hard-line Unionism in Northern Ireland, fulminated against the socialists and deviants who were undermining the British nation from within, and opposed British membership of the EEC. This current despised the social-conservative compromises adopted by Macmillan and Heath. By the late 1970s the old-style nationalists were largely fused with the third major tendency in the party, namely the rising Thatcherite neo-conservatives who, in any case, combined the individualist commitment to economic neo-liberalism with respect for strong, if limited, government and robust nationalism, albeit more pragmatic on Europe than some would have liked.

It has often been pointed out that there is a central tension within contemporary British Conservative thought. On the one hand there is the neo-liberal appeal to acquisitive individualism, aggressive voluntarism, competition, and acceptance of inevitable economic inequality within a free market system which allegedly delivers far greater prosperity than its corporatist or socialist rivals. On the other hand there is the communitarian appeal to traditional moral values and to obedient respect for law, order, and authority in other spheres. Whether or not these two sets of imperatives can be combined coherently in logic or in practice, both are considered necessary to provide a stable, cohesive but economically dynamic social environment (Levitas, 1986; King, 1987; Hayes, 1994). Both sets of beliefs feed into Euroscepticism. On the one hand in the economic and social spheres Conservative Eurosceptics are repelled by what they regard as the interventionist, broadly social-democratic ethos of the EU, with its statist concern for regulation and harmonization, its interference in markets via the CAP and other devices, its redistributive uses of funds from budgetary contributions, its economic committees and its encouragement of corporatism, or its interest in employment security, benefits and entitlements of workers. All of these tendencies are perceived as not only detracting from economic competitiveness, but as also feeding the infernal cycle of the dependency culture whereby member states, regions, and interest groups of all sorts become accustomed to relying on the central institutions of the European Union for direction and support. These are provided at the price of increased levies from member states, which in turn depend more heavily on the Union and less on their individual resources.

On the other hand the EU's supranational aspects also offend against the Conservatives' attachment to traditional values and norms, because these norms are taken to be specific to British culture. They have to be inculcated by family, schooling and by national authorities if they are to have any real meaning. They are not universalistic abstractions, but practices which emerge from and feed back into a native culture which is the product of the cumulative processes of a particular national history. Conservative Eurosceptics assume that the EU cannot provide this type of framework, since it does not embody a common culture or appropriate values, yet its practice and its rhetoric point towards an abstract, homogenizing conception of European society tinged with residual socioeconomic egalitarianism. More generally, Conservative Eurosceptics hold that the EU substitutes arcane formal procedures and bureaucratic regulation for the informal, interpersonal networks of communication which sustain social bonds. In these respects, therefore, the EU is inimical to the survival of Britain as a national community, just as it also undermines British sovereignty in the political, the economic and the juridical spheres. Beyond this, of course, lies the wider affront to Conservative post-imperial nationalism through the EU's aspiration to incorporate foreign and defence policy increasingly under its auspices, thus depriving Britain of the opportunity to play a distinctive international role by virtue of the accumulated wisdom born of its uniquely rich historical experience.

Other groups around the right-wing fringes of the Conservative Party, shading over to the hinterland between the Conservatives and the extreme right, share many of the same preoccupations to an even more intense degree, especially in their perception of the need for assertive national government and the restoration of social cohesion under traditional moral values deemed characteristic of British nationhood, coupled with vigorous free-market capitalism in the economic sphere. Towards the outer margins of the ideological spectrum, however, the old preoccupation with the corrosive presence of coloured immigrants within Britain, and the need to remove the aliens from the social body, rivals the revulsion against the draining of British identity and autonomy from outside by the no less alien entity that is the EU (for example, BNP, 1997; New Democrats, 1997; Third Way, 1997). On the extreme right the need to sustain and defend the nation in all its realms of activity includes radical agendas of political restructuring (towards authoritarianism for the BNP but varying degrees of direct democracy for the New Democrats and Third Way) and economic reforms with advocacy of distributist or other anti-liberal recipes for rebuilding national community. Implicitly in the case of Third Way, and explicitly in the case of the BNP, this is matched by protectionism so that Britain can learn self-reliance once again and rebuild its own independent productive capacity, rather than relying on foreign investment and imports. In fact, unlike the free-trading Conservatives or even the UKIP and many of the right-wing pressure groups, the full-blooded nationalist position involves rejection of the whole post-Cold War, US-dominated international order in the sense of refusing the WTO in the economic sphere and NATO in the area of defence.

Ideological paradoxes

British right-wing Euroscepticism contains a number of paradoxes. One of these arises from the fact that federalism in the EU is often equated by Eurosceptics with authoritarian centralism at the opposite pole from democracy and pluralism. Genuine theoretical and practical discussion of the nature of federalism does not usually occur. The notion of a federal EU is normally reduced to a simple aggregation of central power within the EU and a consequent subtraction from British sovereignty and parliamentary democracy. Yet, Conservative Eurosceptics are members of a party which opposes federalism – and even campaigned vigorously against more limited forms of regional devolution – within the UK on the grounds that it would lead to the disintegration of Britain. The potential comparison between support for a unitary UK achieved by non-democratic means in earlier centuries and a pluralist, federal EU achieved by voluntary pooling of sovereignty has led to all sorts of evasive claims as to how beneficial and effective the British system is (for example, in Howard, 1996). Only the New Democrats and Third Way on the extreme right escape this contradiction, since they argue against centralism and in favour of devolved government within the UK.

The charge that the EU is undemocratic also raises contradictions and circular arguments. Britain is constantly held up as the cradle and eternal defender of parliamentary democracy. The EP is denounced as an incoherent, impotent talking-shop. But right-wing Eurosceptics are only supportive of democracy when it is based on the nation-state (see, for example, Hague, 1998; Portillo, 1998). Or in the extreme case of the BNP, they do not really support democracy at all. Against the assertion that more majority voting in the Council would be compatible with democratization of the EU, it is argued that this is an infringement of parliamentary sovereignty and is therefore unacceptable. Similarly, respect for democracy could perfectly well lead Britain to endorse greater powers for the EP, but it is argued that the British people want their democracy closer to home. As regards the charge that the national parties represented in the EP are too varied to produce an effective parliamentary assembly, Britain's first-past-the-post electoral system for EP elections was one of the obstacles to the formation of more cohesive, more representative trans-EU political formations on the basis of common lists under a shared PR system. Yet, Conservative Eurosceptics argued – albeit with some equivocal shifts of ground in the debates of 1998 over Labour's proposed system of PR by closed lists – that this was the British way, and if discarded for EP elections it would undermine the sanctity of the existing system for Westminster elections, which were the really important ones.

On a more general point, Conservative Eurosceptics have to reconcile their defence of the nation as the optimal unit of international society with their ideological commitment to free trade in an era of transnational company structures, outsourcing and instantaneous international transfers of capital which undermine the power of national governments over their own economies. Those parties of the extreme right which hanker for national protectionism outside the EU are more consistent in their nationalism, but even more utopian under the circumstances. The Conservatives' inconsistency with regard to free trade is all the more striking alongside their right wing's attachment to highly restrictive immigration controls to protect British labour markets and to conserve the integrity of British culture: the principle of free movement of economic products is not taken to entail the free movement of human producers.

To conclude this part of the discussion, it should be added that British Eurosceptical discourse concentrates far more heavily on what is wrong with the EU than on proposing alternative models of organization and strategies for achieving them. However, at the cusp of Euroscepticism and Euroenthusiasm, in keeping with its claim to be flexible and constructive at the heart of Europe, 'a strong and respected partner' (Conservative Party, 1992, p. 3), the now defunct Majorite compromise ventured a vision of the desired shape of Europe for the future, albeit in very general terms. A Conservative Research Department paper captures the mixture of positive and negative goals in this approach to the EU:

> We have always recognised the importance of fighting for British interests in Europe and of developing a Europe which the citizens of this country can support. By this we mean an enlarged, decentralised, deregulated and competitive Europe – a Europe based on the principles of free trade – not some bureaucratic superstate. We want a Europe of nations working together within a flexible framework which takes full account of Conservative philosophy – liberty under the rule of law; free trade and free market; social responsibility and human rights; sustainable environmental development; and wholehearted backing for the Atlantic Alliance.
>
> (Conservative Research Department, 1995, p. 336)

Exponents of the more radical Maastricht-revisionist standpoint would, by and large, have agreed with the Majorite conception of an intergovernmentalist, free-market, free-trading, deregulated, decentralized, enlarged Europe of nations, while arguing that this was not achievable within the framework set in place by the TEU (see, for example, Howard, 1997; Redwood, 1997; Spicer, 1997). But most exponents have shown little temptation to dwell in detail on a blueprint which they would no doubt consider utopian. The stance is far more one of damage limitation, stemming from a supposition that British involvement in the EU is a necessity in order to have full access to the single market. Margaret Thatcher is a partial exception insofar as her notorious speech on 20 September 1988 at the College of Europe in Bruges (reprinted in Holmes, 1996, pp. 88–96) is largely devoted to proposing a set of general principles on the basis of which Europe ought to be constructed. As for the EU-rejectionists, they too tend to be vague on the role of Britain as an international actor outside the EU. The argument is essentially that, whether or not Britain retained some form of association with the EU, it would not be significantly disadvantaged, given the low-tariff world created by GATT. Britain can still trade with EU countries under the GATT provisions, as other non-members do. Major companies will still have access to EU markets from Britain and continue to enjoy Britain's flexible, competitive conditions of production, so there is no reason for them to disinvest. More generally, it can be claimed that if other countries, such as Norway, Switzerland, Canada or Japan, can prosper as free agents, there is no reason why Britain should not do likewise, since there is no correlation between size and wealth (see, for example, Jamieson, 1997; Lamont, 1995, pp. 19–33; UKIP, 1996).

Eurosceptical mythology

Like all ideologies, Euroscepticism has been a spur to myth-making. By myth-making I do not mean production of stories which are necessarily false, but narratives which are ideologically coloured in their selection and/or interpretation of past, present and predicted events, whether or not the facts recounted are more or less accurate (Flood, 1996). There are two principal types of political/historical accounts generated by British Eurosceptics – though the two types are not mutually exclusive, so both may appear in the same discourse.

The first major type of myth involves explanations in terms of conspiracy. There are two main variants, each of which is often framed in the vocabulary of warfare and relentless struggle. One subtype views the history of the EU as a plot by Eurocrats and/or European political elites and/or France and Germany, aided by collaborators in Britain, to deprive Britain of its independence. There is a pessimistic assumption that all non-British actors in the affairs of the EU (or any other political field for that matter) are driven by self-interested pursuit of power, often lubricated with corruption. Britain was knowingly tricked by Edward Heath and his acolytes in league with continental Euro-integrationists into joining an organization which had a hidden federalist agenda. More recently, the heightened pace of integration under the presidency of Delors – the man with the 'tight, unsmiling little bureaucrat's face', as Booker and North describe him (1996, pp. 231–2) – was another instance of stealth, as the real objective of creating a federal superstate has constantly been masked. The frequent attribution of base, cynical motives to ideological enemies, who are never credited with an ounce of vision or idealism, is exemplified in the charges made by Bernard Connolly – himself a former senior official in the EMS – on the subject of the ERM:

> The villains of the story – some more culpable than others – are bureaucrats and self-aggrandizing politicians. The ERM is a mechanism for subordinating the economic welfare, democratic rights and national freedom of citizens of European countries to the will of the political and bureaucratic elites whose power-lust, cynicism and delusions underlie the actions of the vast majority of those who now strive to create a European superstate. The ERM has been their chosen instrument, and they have used it cleverly.
> (1996, p. xx)

A more extreme expression of the conspiracy mythology is the thesis which tries to establish ideological continuity between the EU and pre-war Nazi notions of the New European Order (Atkinson, 1996; Faiers, 1996; Laughland, 1997; and Connolly, 1996, for a weaker version). At its most paranoid it has picked up some of the obsessions of the American extreme right in attempting to imply that the project is aimed towards a new totalitarian tyranny extending to the whole world, as the EU interlocks with the globalist dreams of the secretive participants in the international Bilderberg conferences (including Heath, Carrington and Blair), the Council on Foreign Relations and other related groups of plutocrats and political fixers seeking to create the New World Order. According to Rodney Atkinson, president of the Campaign for UK Conservatism, and to Norris McWhirter, president of the Freedom Association, it was the Bilderbergers who plotted the fall of Thatcher in every detail because she stood in the way of the federalist project (Atkinson and McWhirter, 1995; Atkinson, 1996).

The second major class of narratives corresponds to variant forms of traditional nationalistic, often xenophobic, mythology based on deterministic notions of a permanent, transhistorical British national character conferring a particular direction and patterns of behaviour. This is often taken to be symbolized by figures such as Pitt (with echoes of struggle against French tyranny) or Churchill (with implicit echoes of struggle against German tyranny). Britain is represented in the light of a glorious history of past grandeur which is taken as testimony to its need to be a free, independent actor with the whole world as its field, engaging in cooperative arrangements on a pragmatic basis where useful, fighting the good fight (often alone) against despotism. In contrast, continental Europe is seen as other, incapable of ordering itself, and with its two major powers suffering from serious defects of national character – Germany drawn always to try to dominate but ultimately clumsy and self-defeating, France selfish, untrustworthy, cunning but ultimately lacking in grit. Much of the literature is haunted by the ghosts of World War II. Bernard Connolly (1996), for example, makes frequent references to the Vichy regime's collaboration with Nazi Germany during the Occupation, and uses it as an analogy with the present-day conduct of the two countries. In the EU, despite the covert rivalry between them, France and Germany dream of dominating a new Holy Roman Empire with countries such as Britain tamed on its periphery. Britain, of course, with its transhistorical essence of courage, freedom, democracy, independence and openness to the world, has previously avoided the follies of continental powers and has saved them from themselves in two world wars. Despite the dangers threatening it today, Britain is fitted to show the way towards a more suitable European order.

On the other side of the conspiracy mythology, faced with the duplicity of the Continentals and Brussels, Britain is often portrayed as an innocent. Whereas Eurocrats and their Continental sponsors are viewed as being entirely motivated by self-interested calculation, Eurosceptical British politicians are represented as altruistic defenders of sacred principles, such as 'our democracy' or of the conditions necessary to a dynamic, modern capitalist system. Thatcher is the martyr figure to which Conservative writers constantly return, since she is represented as having regenerated Britain during the 1980s before being brought down by Europhile plotters (for example, in Booker and North, 1996, p. 212; Gorman, 1993, pp. 3–28). Part of this mythology is based on a view of Britain as an embodiment of honesty and integrity, hence easily fooled by Continental deviousness. Thus, Britain is the state which enforces EU directives, etc., while the cynical Continentals merely treat them as window-dressing. We play by the rules.

Where the moderates and the revisionists assume Britain's exemplary role to be one of solitary, heroic resistance to the follies of Brussels until the other member states see sense, Europhobic rejectionists view British withdrawal as a liberation. Their version of the second class of myth can include a utopian vision of the future, conditional on escaping the tentacular clutches of Brussels to enter a new era of freedom and renewal, aided by the vast sums of money saved by no longer having to pay the EU's exactions. Britain can now look after its own people at home and look outwards to the wider world as a free-trading actor. Thus, Britain's productive capacity and trade with other areas of the world flourishes as a result of the dynamic impulse of freedom regained. Having recovered its sovereignty and its independence, Britain is free to enjoy its own unique political culture, its social harmony, its vibrant, entrepreneurial economy and its own way of life. Its example would show the way to the rest of Europe. Christopher Booker and Richard North rhapsodize as follows:

> When we finally summon up the courage and the will to leave this growing catastrophe, this failed mess of a sentimental dream

which, by its own self-contradictions, is inevitably doomed to fall apart in catastrophe, we shall find that it is a truly liberating moment. We may even find it timely to rephrase those immortal words of William Pitt the Younger after Trafalgar: if Britain can save herself by her exertions, she may yet save Europe by her example. Along with our recovered self-respect, it will release a great charge of national energy. We shall once again be able to walk tall and free. It will be the finest thing this country has done since we helped lead Europe to victory over tyranny in 1945.

(1996, p. 237)

Instrumental aspects

What positive functions does Euroscepticism have for its adherents? Firstly, as a form of nationalism, it can make an appeal to social cohesion. Factors which could undermine the legitimacy of right-wing claims to gain or retain political power can be countered by the invocation of national strength and expansion in the world or/and the call for defence against external and internal threats. In Britain many Conservative Eurosceptics would deny that they were nationalists because they do not like the historical connotations of the term. For example, Margaret Thatcher turns the charge against the EU, when she claims that it is 'an obstacle to fruitful *internationalism*'. As evidence of this she cites its attempts to sabotage the GATT negotiations, its trade disputes with the US, the trade barriers erected against Eastern Europe and the EU's undermining of NATO with the absurd project for a defence pillar in its divisive quest to become a superpower (Thatcher, 1996). Thus, in Thatcher's terms, 'the European federalists are in fact "narrow internationalists", "little Europeans" who consistently place the interests of the [European] Community above the common interests of the international community' (1996).

Nevertheless, the Conservative right has clearly made considerable play with nationalism, despite or because of its attachment to an economic creed which accepts the full force of globalization in the name of international free trade and opposition to legislation, regulation or any other 'artificial' obstacle placed in its way. Aside from the effects of worldwide developments in capitalism and communications technology, such as downsizing and outsourcing, breakdown of traditional labour markets and career structures, demands for reskilling, etc., the Conservatives from 1979 to 1997 enacted a gradual, anti-egalitarian socioeconomic revolution which produced major relative costs on the poorer sections of society while enhancing the wealth of other sections. At the same time, given the social costs, the whole legitimacy of the Conservative project rested on claims that these policies had been wonderfully successful in economic terms by transforming Britain from the sick man of Europe to a dynamic global actor. However, those claims were themselves challenged by critics in the war of statistics.

That being the case, the preservation of the nation-state, considered as a natural basis for community, identity and government (against EU power-grabbing, bureaucracy and economic inefficiency) offered a distraction from social division, a substitute for other forms of social solidarity, a call for protection of the Conservative socioeconomic model against further EU encroachments and a channel for displacing the responsibility for unfulfilled promises onto the EU itself. The need to defend the nation against all manner of threats allows appeals to strong leadership and unswerving obedience. The imperative of defending British parliamentary democracy as a beacon of political practice permits Conservative MPs to reject an alternative which would further reduce their professional status as legislators. Conversely, ideological enemies can be denounced for having insufficient care for the interests of the nation, hence being at least indirectly at the service of the forces which are undermining it or preventing it from asserting its legitimate role in the world. Thus, until a few years ago, Western democratic socialists were often accused of being unwitting or conscious pawns of international communism radiating out from Moscow. Today, right-wing Eurosceptics habitually accuse Labour, the Liberal Democrats and the Europhile wing of the Conservative Party of being pawns of Brussels in its relentless struggle to pursue integration at the expense of British sovereignty, independence and economic competitiveness.

Beyond the larger strategic uses of nationalist ideology in public debate there is the basic tactical use of Euroscepticism as a mobilizing device for electioneering purposes and/or in the service of competition for power within the Conservative Party. In the past, Conservatives have been accused of playing the race/immigration card for similar purposes. As was seen in the long run-up to the 1997 election campaign, and in the campaign itself, Eurosceptical discourse plays well with substantial sections of the electorate. For example, a Gallup poll published in the *Daily Telegraph* on 16 November 1996 showed that, while a clear majority believed Britain should stay in the EU – 47 per cent to 33 per cent – only 31 per cent thought Britain's membership was positively a good thing, while 30 per cent considered it a bad thing, 23 per cent neither good nor bad, and 16 per cent with no opinion – the lowest positive support since 1987. Although it did not serve well enough to outweigh negative perceptions of the Conservative Party's governmental record and likely future performance, it is not impossible that under the leadership of William Hague or a successor, with most of the 1997 intake of MPs vigorously Eurosceptical, the party will move further towards nationalist and populist positions coupled with a radical right-wing social agenda heavily influenced by the Republican right in the US (as predicted in *Searchlight*, 1997). This would entail further subjugation of the residual Europhile left of the party, assuming there was not a split. It would also offer the prospect of rallying support from those who had formerly been drawn to the minor Europhobic organizations, leaving the extreme right even more marginalized than at present. But, of course, nothing is certain in these matters.

Conclusion

Euroscepticism does not have to be objectively 'right' or even internally coherent to be viable as an ideology. It merely needs to be coherent enough to satisfy believers, and most people – even including the theoretically minded – will accept gaps and inconsistencies, if they are aware of them, on the ground that they will sooner or later be resolved by someone else. Conversely, a critique does not have to be accepted, even if it is rigorous, so long as it can be rejected as biased by ideological hostility. It is always possible to believe that, because the ideas are valid, they will eventually be understood by a sufficiently wide audience to bring them into the range of practicability. As an ideological configuration, Euroscepticism has produced detailed arguments which add up to a systematic critique of the institutions, the practices and the principles which constitute the EU. Whether or not they are formulated coherently in any given text by a particular individual, and regardless of whether one accepts or rejects these arguments, they are not inherently less rational than those produced by supporters of integration. The alternatives to integration offered by revisionists or rejectionists are not necessarily more utopian than the federalist objectives of Europhiles.

Euroscepticism offers a potent ideological channel for various forms and degrees of right-wing nationalism. It is driven in large measure by the claim that Britain is losing its very essence as a state and as a culture by being absorbed into, and marginalized by, the EU. The corollary response is to make a virtue of marginality, while asserting Britain's need, and historical right, to play a pivotal role in the wider world outside the claustrophobic confines of Europe.

Notes

1 For convenience of nomenclature the term 'European Union' and the acronym 'EU' are used throughout this chapter to refer to the organization in all of its historical stages, including European Economic Community and European Community.
2 For representative collections of predominantly Conservative Eurosceptical sources, see Hill, 1993; Holmes, 1996.

References

Atkinson, R. (1996) *Europe's Full Circle: Corporate Elites and the New Fascism* (Newcastle: Compuprint).

Atkinson, R. and McWhirter, N. (1995) *Treason at Maastricht: The Destruction of the Nation State*, 2nd edn (Newcastle: Compuprint).

Beloff, Lord (1996) *Britain and European Union: Dialogue of the Deaf* (Basingstoke: Macmillan).

Booker, C. and North, R. (1996) *The Castle of Lies: Why Britain Must Get out of Europe* (London: Duckworth).

British National Party (1997) *Britain National Party Election Manifesto 1997* (http://ngwwmall.com/frontier/bnp/manint.htm).

Cash, W. (1992) *Europe: The Crunch* (London: Duckworth).

Connolly, B. (1996) *The Rotten Heart of Europe* (London: Faber & Faber).

Conservative Party (1992) *The Best Future for Britain: The Conservative Manifesto 1992* (London: Conservative Central Office).

Conservative Research Department (1995) *Europe: The Right Approach* (London: Conservative Research Department).

Faiers, R. (1996) 'The Betrayal of Britain', *This England*, 29.4 (Winter), 50–2.

Flood, C. (1996) *Political Myth: A Theoretical Introduction* (New York: Garland).

FCO (Foreign and Commonwealth Office) (1996) White Paper on the 1996 Intergovernmental Conference, *A Partnership of Nations: The British Approach to the European Union IGC 1996*, 12 March, (http://www.fco.gov.uk/europe/igc.html).

George, S. (1994) *An Awkward Partner: Britain in the European Community*, 2nd edn (Oxford: Oxford University Press).

Goldsmith, Sir J. (1994) *The Trap* (Basingstoke: Macmillan).

Goldsmith, Sir J. (1996a) *Sleepwalking into the European Superstate* (London: Referendum Party).
Goldsmith, Sir J. (1996b) 'The Betrayal of Our Nation', speech to the Referendum Party Conference, Brighton, 19 October 1996 (http://www.referendum.org.uk/keynote/boon2.html).
Gorman, T. with Kirby, H. (1993) *The Bastards: Dirty Tricks and the Challenge to Europe* (London: Pan).
Hague, W. (1997) 'Extracts from a Speech by the Rt Hon William Hague MP, Leader of the Conservative and Unionist Party, Addressing the 1997 Scottish Conservative and Unionist Conference at the Dewars Rink, Perth, on Friday 27th June 1997 at 1645 Hours', press release (http://www.conservative-party.org.uk/newspags/1088-97.htm).
Hague, W. (1998) 'The Potential for Europe and the Limits to Union', speech at INSEAD Business School, Fontainebleau, 19 May (http://www.conservative-party.org.uk/newspags/05181500.htm).
Hayes, M. (1994) *The New Right in Britain: An Introduction to Theory and Practice* (London: Pluto).
Holmes, M. (ed.) (1996) *The Eurosceptical Reader* (Basingstoke: Macmillan).
Holmes, M. (1998) *John Major and Europe: The Failure of a Policy, 1990–7* (London: Bruges Group Occasional Papers, 28).
Howard, M. (1996) 'Liberty and the Nation', speech to Conservative Way Forward, Wessex Hotel, Bournemouth, 8 October (http://www.conservative-party.org.uk/newspags/617-96.html).
Howard, M. (1997) *The Future of Europe* (London: Centre for Policy Studies).
Jamieson, B. (1997) *Britain: A Global Future* (Oxford: Nelson & Pollard).
Jenkins, L. (1997) *Britain Held Hostage: The Coming Euro-Dictatorship* (Washington: Orange State Press).
King, D. S. (1987) *The New Right: Politics, Markets and Citizenship* (Basingstoke: Macmillan).
Kitzinger, U. (1973) *Diplomacy and Persuasion: How Britain Joined the Common Market* (London: Thames & Hudson).
Lamont, N. (1995) *Sovereign Britain* (London: Duckworth).
Laughland, J. (1997) *The Tainted Source: The Undemocratic Origins of the European Idea* (London: Little, Brown).
Levitas, R. (ed.) (1986) *The Ideology of the New Right* (Cambridge: Polity).
Morris, P. (1996) 'The British Conservative Party', in J. Gaffney (ed.), *Political Parties and the European Union* (London: Routledge) pp. 122–38.
New Democrats (1997) *A Manifesto for Britain* (http://www.netlink.co.uk/users/natdems/manifesto.htm).
Palmer, T. (1997) *Euroslavia* (Brentford: Pallas).
Portillo, M. (1998) *Democratic Values and the Currency*, speech to Institute of Economic Affairs, 14 January (London: Institute of Economic Affairs).
Redwood, J. (1997) *Our Country, Our Currency* (London: Penguin).
Rifkind, M. (1996a) Statement by the Foreign Secretary, Mr Malcolm Rifkind, in the House of Commons, 12 March (http://www.fco.gov.uk/current/1996/mar/12/rifkind/statement/igcwhitepaper.txt).
Rifkind, M. (1996b) Speech by the Foreign Secretary, Mr Malcolm Rifkind, House of Commons, 21 March (http://www.fco.gov.uk/current/1996/mar/21/rifkind/speech/igc.txt).

Roberts, A. (1995) *The Aachen Memorandum* (London: Orion).
Searchlight (1997) 'The New Right: A Growing Danger', unsigned article, *Searchlight*, 25 January, pp. 6–8.
Seliger, M. (1976) *Ideology and Politics* (London: Allen & Unwin).
Spicer, M. (1992) *A Treaty Too Far: A New Policy for Europe* (London: Fourth Estate).
Spicer, M. (1997) 'Variable Geometry in Europe', in J. Gedmin (ed.), *European Integration and American Interests* (Washington, DC: AEI Press), pp. 41–6.
Tebbit, Lord (1996) 'New Consensus – New Schism'. Speech to the Conservative 2000 Foundation, 14 November (http://www.the2000foundation.org.uk/tebbit.htm).
Thatcher, M. (1996) 'Narrow Internationalism', Anti-European Union Server, monthly analysis, August (http://www.soton.ac.uk/~nss194/Analysis.html).
Third Way (1997) *General Election 1997: Manifesto of the Third Way* (http://www.users.dircon.co.uk/~thirdway/files/3 w97mfst.html).
UK Independence Party (UKIP) (1996) 'UK Independence Party'. Unofficial UKIP page (http://www.bath.ac.uk/~ce5krs/ukip.htm).
Whiteley, P., Seyd, P. and Richardson, J. (1994) *True Blues: The Politics of Conservative Party Membership* (Oxford: Clarendon Press).

10

Home at Last? Czech Views of Joining the European Union[1]

Peter Bugge

Czech attitudes to the European Union form a puzzling picture. On the one hand, the Czech wish to join the EU has been unequivocal since 1989, as has the Czech insistence that the nation is ready for membership and has a right to it. But on the other hand, actual policies towards the EU have been characterized by a strange passivity: the country was the last but one among the Central and East European nations to apply for membership (only Slovenia came later), and domestic preparations have been scarce and often late in coming, both administratively and in terms of informing the public about the consequences of joining the EU.

It may be argued that these contradictions reflect a tension in the Czech discourse on these matters between a claim to *belong* to Europe and a call for a *return* to Europe (or, in a similar vein, for a *Europeanization* of the country). This tension may be found in many different contexts and it deeply affects the perception of what integration means, and hence what it implies for the Czech nation. If the Czechs already *are* a legitimate and integral part of Europe, it becomes Europe's responsibility to recognize this fact and adjust to it, while the main Czech task consists in making this point. But if the wish to *become* (again) European is stressed, the focus shifts to the Czechs themselves, who then have to move from their periphery 'towards Europe'. Here we shall follow some expressions of this tension in official policies, in the discourse on Europe of key political actors, in academic discussions of Czech foreign politics, and finally in the broader public debate on Czech–EU relations.

Czech policies towards the EU

One may discern four phases in Czech(-oslovak) policies towards the EC/EU since 1989. In the first phase, which lasted until the parliamentary elections of June 1992, the new, mostly ex-dissident leadership quite naturally focused first on emancipating their country from the Soviet grip, i.e. to negotiate a withdrawal of Soviet troops from Czechoslovak territory and a dismantling of the Warsaw Pact and the CMEA (COMECON). As to alternatives, Vladimír Handl has described the foreign policy philosophy of the Czechoslovak government in these years as *'liberal institutionalism'* with a strong belief in the 'cooperative and the institutional potential of international politics' (Handl, 1995, p. 130ff), even if the initial, rather utopian, vision of a new cooperating Europe based on the CSCE and the Council of Europe (and without the pre-1989 structures) was soon replaced by a recognition of the importance of NATO and the EC as cornerstones of the new European architecture (Břach, 1992; Šedivý, 1994/95; Šedivý, 1997).

Inevitably the EC was not the new government's first concern, although Prime Minister Marián Čalfa sent a letter to Jacques Delors asking for talks about 'possible forms of affiliation to the European Community' (quoted from Břach, 1992, p. 123). After a cool reply from Brussels Czechoslovakia concentrated instead on catching up and on coordinating its policies with Poland and Hungary, who as reforming countries before November 1989 had a lead in trade and other agreements with the EC. From August 1990 these 'Visegrád countries' held negotiations with the EC, resulting in the so-called Europe Agreement of December 1991, a new kind of Association Agreement between the EC and each of the three. Czechoslovak hopes of access to markets and membership were, however, far from met in the Agreement: membership was mentioned only in the Preamble – and only as a Czechoslovak *objective*, and the EC excluded a number of sensitive goods from the promised opening of its markets to Czechoslovak exports. On the other hand, the three Visegrád countries committed themselves to adjust their legal systems to the *acquis communautaire* of the EC.

In this way the Europe Agreement demonstrated the weakness and marginality of the Visegrád countries, both in reality and in EC perceptions: the EC level of commitment was low and nearly the whole burden of adaptation to the new European situation was placed on the Visegrád three. Still, attempts to lessen Czechoslovak interest in membership by suggesting alternative, non-committal pan-European structures – as in Mitterrand's 1991 proposal for a 'European Confederation' – were rather bluntly dismissed. In President Havel's words:

> I think that the idea of a pan-European confederation cannot just overlook the existence of the European Community or take this to be something parallel with, and unconnected to, Europe as a whole, to be some kind of exclusive club in which the others have nothing to do. Instead it must understand the European Community as its driving force, its flag-bearer, as a model for its own future.
>
> (Havel, 1992, p. 91)

With the election of Václav Klaus as Czech Prime Minister and the division of Czechoslovakia by the end of 1992 a second phase began. The new right-wing government was dominated by economists with no dissident background, and the change of both style and course was tangible. In Handl's words, Czech foreign politics was now marked by '*realism*', i.e. by an emphasis on state-to-state relations and a scepticism of institutionalized multilateral cooperation. All Visegrád collaboration was abandoned, as the Czechs were convinced that they could do better alone. The Klaus cabinet was very optimistic about the country's economic performance, and Klaus proudly proclaimed in 1993 that 'we will be ready to join the EU earlier than the EU will be ready to accept us' (quoted from Handl, 1995, p. 136), while at the same time criticizing the Maastricht Treaty in the spirit of the British conservatives.

The government's foreign policy statement of April 1993 reflected these changes. To secure the stability, security and economic prosperity of the country, it was said, the Czech Republic had to create good relations with its neighbours and to work on its own gradual integration into the 'main European economic, political and defence organizations'. This ranking of neighbours above international organizations was echoed in an account of bilateral and multilateral priorities with a clear emphasis on the former. And in respect of the latter, Foreign Minister Joséf Zieleniec stressed the significance of integrating into the *world* economy before specifying priorities *vis-à-vis* the EC. EC membership was defined as a 'long-term goal and a clear priority', and to prepare for a 'fast and smooth' accession the Czech Republic would have to harmonize its legislation and create a free-trade zone. Integration was, the government acknowledged, first of all a domestic challenge. But:

> on the other hand, the fall of the Iron Curtain is not a challenge only to us. The nations of Western Europe also have gradually to get used to the thought that we will one day live together. The European Community, to a large extent a product of the former bipolar world, must find its new place in the new world, must find its new vision.
>
> An effort to adapt the economy must be made on both sides of the former Iron Curtain. It is obvious that, as a consequence of the integration of our countries, even the economy of the countries of the European Community must undergo a process of restructuring, which is also not going to be painless.
>
> (Zieleniec, 1993, pp. 243, 249)

Enlargement was thus presented as making basically equal demands on equal partners, and the government felt confident enough to tell the EC how to behave, while taking for granted its willingness to enlarge.

Meanwhile, the government paid little attention to domestic legal or administrative preparations for EU membership. An October 1991 government decision decreed that all draft laws should aim at compatibility with EC law, but little happened in the following years. This was to a large extent a consequence of the government's purely economic approach to the transformation of Czech society, which led to a neglect of law as such and thus of the consistency of the legal framework. In addition, few resources were devoted to development or restructuring of public administration, and no special government body was created to deal with EC/EU matters (Desný, 1997, p. 45f.).

Nor did the EU hasten to live up to Czech expectations. At one point Hans van den Broek had to tell Klaus that 'it is not the European Union which wants to join the Czech Republic' (quoted from Rhodes, 1999, p. 67, note 41). After Czechoslovakia was split into two states a new Europe Agreement had to be negotiated, which in some respects turned out to be less favourable to the Czech Republic than the previous one (Handl, 1995, p. 136f). Although the June 1993 Copenhagen summit, by recognizing their wish for membership, brought a breakthrough for the Central and Eastern European countries, only very general membership terms were laid down, with no timetable. Czech–EU relations were thus characterized by a certain immobility. One critical observer described the Czech government's attitude as *'Euro-passivity'* (Jakš, 1994, p. 148).

In 1995 a third phase set in, as the government began to moderate its self-confident, 'Euro-sceptical' rhetoric, and to revise its EU policies. In January a Government Committee for European Integration was created, headed by the prime minister and assisted by a Working Committee and numerous Working Groups. New initiatives were taken to accelerate the process of legal approximation, including an increased use of legal assistance from the PHARE programme (Zemánek *et al.*, 1997, p. 155f.; Desný, 1997, p. 47f). Finally, on 23 January 1996 the government submitted its application for EU membership, accompanied by an explanatory Memorandum.

The Memorandum was very frank about the government's difficulties in overcoming its mistrust towards the EU, and giving up part of the country's sovereignty. In the end, the text said, the government had however concluded that the situation 'allows for no alternatives'. With a marked change of tone the Memorandum then states that 'the Czech Republic accepts for its future membership the European Union as it is and as it will be shaped by the collective wisdom of its Member States in the months and years to come', and that 'the Czech Government also accepts the broader, non-economic aspects of European integration' (*Memorandum*, 1996, p. 260).[2] Similarly, after the June 1996 elections, which forced Klaus to form a minority government, new government policy statements included the fastest possible membership of the EU and NATO as key government priorities, before any mention of bilateral relations or world trade and the economy (documents in Kotyk, 1997, pp. 241–2 and 251–7).

The EU also began to act. The May 1995 White Paper finally gave specific suggestions for how to prepare for the Single Market to include the Central and Eastern European countries, and at the Madrid summit of December 1995 it was decided that accession talks were to begin six months after the conclusion of the Intergovernmental Conference. This process led to the Commission's *Agenda 2000*, a comprehensive evaluation of all applicant countries presented in July 1997, and then to the December 1997 decision to start negotiations with the Czech Republic and five other countries. While new activism on both sides brought the Czech Republic closer to the EU, Czech commitment to preparations and adjustments at home remained half-hearted. The Government Committee met only once or twice a year. *Agenda 2000* described the attitude of the Klaus government:

> Confident in its progress towards meeting the obligations of EU membership, the Czech Republic has at times shown signs of reluctance to acknowledge difficulties and seek a collaborative approach to resolving them.
>
> (*Agenda 2000*, p. 8)

By sheer coincidence the fall of the Klaus government in November 1997 came almost at the very moment of the decision to enlarge. So, a fourth phase may be said to have begun with the appointment of the Tošovský government and the formal launch of the enlargement negotiations in March 1998. Tošovský's cabinet was partly a caretaker government with a limited mandate, but the government's policy statement from January 1998 revealed an 'activist' stance in the EU question. The government promised to create a programme of national preparation for the EU accession talks before March, and brought up the EU twice in the domestic policy part of the statement in undertakings to step up legal harmonization and adopt regional policy principles in keeping with European Union practice (*Policy statement*, 27 January 1998). Likewise, the new Foreign Minister Jaroslav Šedivý expressed a philosophy of liberal institutionalism, while promising a more clearly asserted '*European dimension*' in Czech foreign policy, including a new emphasis on regional cooperation (Šedivý, 1998, pp. 6–9).

These views of the Tošovský cabinet were close to the official position of the Czech Social Democrats, who after their election victory in June 1998 formed a minority government. In its years in opposition the party presented itself as the most genuine adherent of the EU, claiming that it, unlike Klaus, welcomed *all* dimensions of the EU project. This was stressed in the party's 1996 election programme, which defined the EU 'not just as a zone of free trade, but as a multidimensional European community, united by a common social, ecological, agricultural, transport, regional, foreign and security policy'. The election programme from April 1998 promised to step up integration efforts at all levels.

The Social Democrat government policy statement from August 1998 also abounded with promises to bring Czech legislation and political practice into line with EU standards. The government was very critical of the present position of the country, both *vis-à-vis* its neighbours and in Europe at large. It demanded a 'comprehensive inventory of the state of Czech society', which should 'map the scope of our internal debt and define the distance separating us from the average standard of European Union countries' (*Policy statement*, August 1998, Chapter 3). In the same vein, the government expressed its conviction that

> our future integration into the European Union will help Czech society overcome some of the negative attitudes to foreign-speaking, looking and living people which are a hallmark of all isolated communities. The Government will do everything possible to make Czech society open itself up to the greatest possible extent to Europe and the world and transform itself into a multicultural society.
>
> (Ibid., Chapter 4.1)

One may wonder to what extent the Social Democratic electorate would subscribe to this; but the image of the Czech Republic as an 'isolated community' faced with the task of overcoming the 'distance' to the EU and thus to 'European' standards is very clear.

Still, at the time of writing, the performance of the government during its first months in office has not been too impressive. Many legislative initiatives are being prepared, but actual progress is slow, and certain party policies, both before and after coming to power, have pointed in other directions. The Social Democrats seem very reluctant to go ahead with the privatization of banks, as demanded by the EU, and populist anti-German sentiments have at times been voiced. In the European Commission's November 1998 Progress Report the Czech Republic was strongly criticized for having made very little progress in adopting the *acquis*. The government has largely accepted this verdict; but it has put all the blame on the previous governments and insisted that the EU takes a favourable view of present Social Democratic plans and policies. Thus, it remains to be seen if the Social Democrats are able to transform their verbal commitment to integration into coherent, functioning policies.

Talking Europe: Havel and Klaus on Europe and the EU

Inevitably, Czech politicians have sought to embed their actual decision-making in a broader explanatory framework. For political power rests not only on offices, but also on the ability to define the agenda and direct public discourse. The Social Democrats have come to power with a new vision of Czech–EU relations, but one can well argue that until 1997 the Czech political scene – especially at the level of discourse formation – was dominated by two actors, President Václav Havel and Prime Minister Václav Klaus. The two often disagreed on Europe and the nation's place in it, but underneath one detects some shared assumptions that may be said to form the core of the Czech world view in these years. In brief, the 'Czech view' was that the country was no outsider in the first place: it had belonged to Europe for centuries and was simply returning to its natural home.

Before 1989, foreign politics was not one of Václav Havel's main interests. But as president he has formulated a comprehensive vision of a united Europe, based on an enlarged EU. This new community, he wrote in 1991, should be based on the principle of 'unity in diversity'. He evoked a future Europe where the EC had a strongly integrated political leadership, where borders had become a mere formality, and where many legal or other competences were delegated up to a supranational or down to a regional level (Havel, 1991, pp. 63 and 87). In Havel's view this political goal was a natural consequence of Europe's being *one civilization*, increasingly referred to as 'Euro-Atlantic' or 'Euro-American' – while Russia, by contrast, was a 'Euro-Asian' power. Havel evolved a whole philosophy of European history.

Europe was not just a civilizational and cultural entity, he claimed: 'Europe has always been and still is in essence one single and indivisible political entity, though immensely diverse, multifaceted and intricately structured' (Havel, 1997, p. 56). For Havel, European history has essentially been one long attempt to shape this structure and define the relations between its parts. Hitherto, however, the outcomes have mostly been orders built on force, and consequently overthrown by force. The EU, by contrast, is a 'magnanimous attempt' to give Europe an order based on democracy, peace and cooperation: Europe now had a historic opportunity to extend this principle to the whole continent. Havel thus derives from history his country's *right* to EU membership, making political and economic criteria secondary considerations. This makes the obligations on the EU side stronger than those on the applicant side. Even in speaking to a domestic audience, Havel has said little about what the Czechs have to do on their side to prepare for EU membership.

Havel interprets all tensions in Europe after 1989 as a struggle between the democratic, *'civic'* principle (the *idea* behind the EU) and the anti-democratic principle of nationalism and authoritarianism. He vehemently rejects the principle of ethnically defined nation-states and calls European integration the best way to overcome the dangers of nationalism. This would not, he is sure, destroy cultural, ethnic or regional differences in Europe or create a 'monstrous superstate'. Instead, 'the many different civil societies in the democratic European countries will, together, create the great European civil society' (Havel, 1995, p. 14f.). Thus it was that Havel could tell the European Parliament in 1994 that: 'we are able and happy to surrender a portion of our sovereignty in favour of the commonly administered sovereignty of the European Union, because we know it will repay us many times over, as it will all Europeans' (Havel, 1995, p. 58).

Whilst this idealized EU became a key positive actor in Havel's grand narrative of a European battle between good and evil, actual EU performance merited a less favourable evaluation. In the same 1994 speech Havel praised the Maastricht Treaty as a great 'technical' achievement, but added that it left him with a feeling of something lacking. He was convinced that the EU did have a spirit or an ethos, but it had disappeared behind the technicalities and petty arguments about rules and regulations. To win people over, Havel claimed, the EU must 'impress upon millions of European souls an idea, a historical mission and a momentum. It must clearly articulate the values upon which it is founded and which it intends to defend and cultivate. It also must take care to create emblems and symbols, visible bearers of its significance' (Havel, 1995, p. 62): in short, the EU must conduct an active 'identity politics'.[3]

Havel asked from the EU a clear commitment to enlargement, including a timetable for all of Europe not yet included. But if Western European decision-makers hesitated to accept his logic, Havel's cultural–historical line of reasoning left him in a vulnerable position. He has often criticized the reluctance of the West to change. Yet, without concrete, positive *political* arguments for enlarging 'too fast what has been created for decades and what has proved its worth in the free world' (Havel, 1992, p. 88), Havel has had to resort to appeals to a Western 'negative' self-interest, evoking a horror scenario of a nationalistic, chaotic Eastern Europe from which conflicts will inevitably spill over to the West. It would be naive and detrimental, he has insisted,

> to believe that one half of Europe will blossom, will be able to protect itself from different dangers and cooperate along democratic principles and that the other half of Europe will forever find itself in some indeterminate vacuum. . . . It is now six years ago that the Iron Curtain fell. I feel that relatively little has happened in these years. And time is working against the democrats.
>
> (Havel, 1996, p. 79)

Such open threats and appeals were rare in the discourse of Václav Klaus. But behind his self-confident appearance one detects a profound ambivalence in his perception of the Czech position in Europe. Klaus once insisted that 'the Czech lands have always been a part of Europe', only to admit in another context that 'if we want to live in European contexts, after forty years of separation, we must Europeanize ourselves internally and likewise in our foreign politics' (Klaus, 1994, pp. 166 and 132). Similarly, with a critique of the EU's habit of monopolizing the concept of Europe patently in mind, he could say that 'Europe can never be enlarged (or narrowed) by anybody. Europe is Europe no matter which institutions emerge in it' (ibid., pp. 166–7); but at other times he himself succumbs to the same impulse: 'Europe should strive to contribute to the integration and inner stability of the countries of the former Eastern bloc' (ibid., p. 137). Again, explaining his rejection of Visegrád cooperation he wrote that 'we have resisted the recommendation of some of our western friends to form in Central and Eastern Europe some special, sub-regional institution, because this would only separate us from Europe, not unite us with it' (Klaus, 1995, p. 123).

Like Havel, Klaus offers a 'philosophy' of European history. To him, Europe is an entity 'based on a distinct heterogeneity which creates a very fragile balance' (ibid., p. 141). Development then stems from the tension between 'unifying, pan-European tendencies on the one hand and individualizing, more national strivings on the other' (Klaus, 1994, p. 53). Since this conflict is as old as Europe itself, it would be naive to think that it could be overcome in the 1990s. Although Klaus acknowledged that both fragmentation and forced unification were bad for Europe, he has usually warned *against* ideas of EU unification. To the struggle of individuality against unification Klaus has added a list of further oppositions between good and evil: free market versus protectionism, liberalism versus statism, common sense versus dogma, truth versus lies. In this setting he clearly places on the side of the devil the adherents of 'pan-

European government, pan-European currency, pan-European standards for the shape and size of bottles of beer or wine, a common European citizenship, a common European social legislation and such things' (ibid.).

'The whole idea of Europe should not be based on too simplistic a rejection of patriotism and national feeling', Klaus felt (1995, p. 123); and to him the natural unit for self-determination was the sovereign nation-state, the source of all European supranational institutions' authority. Klaus therefore defined the key Czech challenge in the post-1989 era in somewhat defensive terms: 'The Czech Republic is facing one important task: how to be European without at the same time dissolving in Europeanness like a lump of sugar in a cup of coffee' (Klaus, 1994, p. 136). Or again, putting a new twist on the 'return to Europe' metaphor, '[we] stand before a double task: to find our own identity and not to lose it again immediately on our way to Europe' (ibid., p. 153).

Klaus's views on the history of Western European integration seem contradictory. At one time he presented the EC of old as founded on ideological paradigms from the first two-thirds of the twentieth century: a distrust of a spontaneously working free market and a belief in size, unification, planning and state intervention. This was the paradigm behind the Maastricht project; but it appears increasingly inadequate and outdated, both practically (Klaus sees it as a main cause of Western Europe's economic problems and of its protectionist stance *vis-à-vis* Eastern Europe), and theoretically, since the new, liberal paradigms of the 1980s have gained in political impact. Therefore, Klaus predicts, the EU is going to shift towards deregulation and economic openness (ibid., pp. 135, 139, 151 and 164). But Klaus has also written of Maastricht not as the swansong of an old EC philosophy, but as the first omen of a new and threatening attempt to replace the original vision of the EC. The EC originally aimed to prevent a return to war by integrating Germany in a new fashion; to support the values of freedom and democracy against Communism; and to stimulate growth by removing barriers to trade and creating a Common Market. With Maastricht there has appeared a far more comprehensive vision of greater coordination from the centre, a reduction of the authority of the nation-state, joint programmes even in non-economic areas, and efforts to impose a new European identity (Klaus, 1995, p. 144).[4]

This contradiction may reflect a changing perception of the future course of the EU, but it also reveals an ambiguity in Klaus's perception of the value of the EU. In the second version he clearly recognizes the *political* value of the EC/EU project, while elsewhere he denies it any value at all: 'the success of western Europe depended not on the institutions of the European Community, but on a free society, private ownership and a free market' (Klaus, 1994, p. 166).[5] Being so sceptical of the EU, Klaus was rather vague in his arguments as to why his country wanted to join. 'We do not want to miss the advantages that come from the membership of European institutions', he said without specifying these advantages, while adding vaguely that 'we share the European values of our Western neighbours' (Klaus, 1995, p. 121). Nor has Klaus specified *why* he considers enlargement to be 'an enormous and at the same time unrepeatable European chance and challenge' for the EU (Klaus, 1997, p. 1).

Klaus's ambivalent attitude to the EU was revealed again in his reaction to the Commission's November 1998 Progress Report. The report's picture of the situation was wrong, he said, attributing its negative view (in a way that challenged the legitimacy of the whole enlargement process) to a conspiracy against enlargement on the part of bureaucrats. '[C]ivil servants have nothing to blame the governments of any state for', he claimed; the report was part of a 'conscious policy of European bureaucrats. In some countries there is a strong campaign against the enlargement of the EU' (*Lidové noviny*, 5 November 1998, p. 7). So, again, Klaus appears just as afraid of exclusion as he is sceptical of the institution he asks to join.

In sum, Havel and Klaus both use historical and moral arguments for Czech membership rather than concrete political ones. What separates the two is their understanding of the consequences of membership for national identity. For Havel the EU is to protect the Czech nation from its own latent nationalism; the EU as a *solution* to the problem of nationalism and national isolationism. For Klaus the EU is a *threat* to national identity and a legitimate patriotism. Yet in both one detects a fear of rejection and a distrust of actual EU intentions, which testifies to a feeling of Czech powerlessness. Their historical thinking offers divergent versions of 'Europe', whereas the EU's own thinking runs along altogether different, more instrumental lines.

The expert view

Assessing the role of the Czech expert community Václav Kotyk claims that there has been no lack of competent academic experts willing and able to criticize government policies whenever required. The main problem, he insists, rests not with the experts, but with politicians unwilling to listen and to revise even manifestly erroneous policies (Kotyk, 1998, p. 59). Others too have blamed the Czech foreign policy-making process for being highly centralized, with little delegation of tasks and little use of, or dialogue with, experts, social interest groups, or the general public. The outcome has been unrealistic policies and problems of communication (Moorhouse, 1996, pp. 367–71; Handl, 1995, pp. 145–7). Robejšek is equally scathing about the quality of Czech foreign policy-making, but he also finds that the expert community is incompetent and impotent:

> A maximum of twenty people throughout the Republic regularly have something qualified to say about foreign policy. Their discussions take place practically with the exclusion of the public and of politicians, and are characterized by a strange combination of stereotypes and resignation.
>
> (Robejšek, 1998, pp. 78–9)[6]

Some articles, especially from the first years after the partition of Czechoslovakia, could create the impression that there was little for the politicians to listen to, since the authors seemed content with merely echoing government positions and perceptions. Thus, Petr Pavlík gave a very optimistic rendering of Czech integration into European and global economic structures, repeating the Klaus dictum that the Czech Republic would be capable of membership before the EC was prepared to accept the country (Pavlík, 1993, p. 22). In a similarly confident vein Luděk Urban, commenting on the White Paper from 1995, assumed that institutionally and in terms of legal approximation the Czech Republic had done everything necessary to speed up its preparations for EU integration (Urban, 1995, pp. 9–10; see also Winkler, 1996, and Pech and Winkler, 1996 for uncritical accounts of the first years of Czech foreign policy). Most Czech foreign-policy experts did, however, aspire to a higher degree of analytical distance, and certainly many have criticized various government policies. And for our purposes, both the critique and the 'stereotypes' – the basic assumptions underlying all evaluations – are important.

The main point objected to has undoubtedly been the Klaus government's lack of a regional policy. The critics hold that the strategy from 1993 of 'going solo' was wrong, even counterproductive. For the EU saw enlargement in a regional perspective, not in a narrowly national one; and Central European cooperation would enhance the stability of the region and thus increase everybody's chances of getting in. In alienating its neighbours the Czech Republic risked isolation, which might lessen its value in the eyes of Germany and other central EU partners and hamper its own accession. What is more, the cult of isolationism could encourage national egoism and Euroscepticism in the population, which ought rather to be reminded of the value of trans-national cooperation (Leška *et al.*, 1997, pp. 109–12 and 118; Had *et al.*, 1997, pp. 54–5; Handl *et al.*, 1997, p. 169; Kotyk, 1995, p. 71).

The lack of domestic preparations was the second main cause of discontent. Many blamed the government for being too vocal, while not demonstrating the necessary industry in defining its own specific interests and objectives in the integration process. They also objected to the insufficient policy planning, the absence of a genuine cost–benefit analysis of the consequences of membership, and the deficiencies in the sphere of legal approximation (Pick, 1997, p. 13; Kotyk, 1995, p. 69; Desný, 1997). Government statements on the EU rarely showed any deeper understanding of the spirit or letter of the Union, and their critique of Brussels bureaucracy sat akwardly with the call for a speedy integration (Jakš, 1994, p. 145). According to Jakš neither dreaming nor abuse was of any use:

> A rational formulation of Czech interests *vis-à-vis* Europe should, therefore, not proceed from a pro-integration Romanticism divorced from the *acquis communautaire* and based on expectations of essential model changes, or on the hope of concessions motivated by the requirements and problems of Central and Eastern Europe. On the contrary, those interested in membership in the European Union must be the ones to adapt.
>
> (Ibid., p. 148)

Leška makes the interesting observation that, although Czech politicians often referred to 'common values', they did not refer to the values that were cultivated in the EU. Thus, the word 'solidarity' occurred far less in the Czech political rhetoric than in EU statements and programmes; and while all agreed to the values of democracy, human rights and a free market, Czech politicians consistently neglected the welfare and social policy aspects of the EU. This led the government to present a distorted image of what the

EU was about (Leška, 1996, pp. 75–6; Leška *et al.*, 1997, p. 110).

Finally, critics urged the government to start informing the population about the EU and the implications of Czech membership. Without an informed public, the legitimacy of the integration process was endangered, and the population had to be prepared for some controversial aspects of membership – such as the freedom to take up residence in all member states, which in a Czech–(Sudeten–)German context could be perceived negatively (Pick, 1997, p. 16; Had *et al.*, 1997, p. 58; Handl *et al.*, 1997, p. 178).

One finds a total consensus in the Czech foreign-policy literature that EU membership is both advantageous and necessary for the Czech Republic. No 'Euroscepticism' can be found. In the preface to the most comprehensive discussion of Czech foreign politics, *Czech Foreign Politics: Reflections on Priorities*, the editor defines one of the book's purposes as to support the Czech efforts to join the EU and NATO by giving solid foundations to the debate (Kotyk, 1997, p. 5).

At times the advantage of joining the EU is taken too much for granted even to need an explanation.[7] But when the benefits of membership are being discussed, political considerations clearly carry greater weight than economic ones. In *Czech Foreign Politics'* chapter on the EU, the authors list the key advantages as a guarantee of peace and prosperity; enhanced domestic stability, participation in decision-making; and an easing of the geopolitical problem that arises from being placed between Germany and Russia. Specific economic considerations – such as market access, access to funds and programmes, etc. – are brought up only later. It is stressed that many of these advantages will show only in the long term, while in the short run increased competition may be a problem for the Czech economy. The loss of sovereignty is also mentioned as a disadvantage, but the authors feel confident that the principle of subsidiarity will leave sufficient scope for the preservation of a Czech national identity (Had *et al.*, 1997, pp. 49–53). Their conclusion is clear:

> The Union is particularly attractive for small European states, as it gives them a platform to assert their interests in a broader European framework. None of the small European countries have had such an opportunity in their history. To remain outside the EU means to remain in the periphery of the main stream of the present and future European development.
>
> (Ibid., p. 52)

These main arguments are found again and again. Pick stresses the political importance of participating in European decision-making, since economically the Europe Agreement already grants the Czech Republic sufficient possibilities (Pick, 1997, p. 12).

The 'soft security' aspect is also brought up by many, in particular with reference to Germany: i.e. increased domestic and regional social stability, and protection against economic pressures (Eichler, 1997, pp. 77–8; Kotyk, 1995, pp. 67–9). With or without the EU the Czech Republic is becoming part of the German 'economic sphere', it is argued, and the EU is called a 'continental' solution to the 'German problem'. If the Czechs prefer a 'European Germany' to a 'German Europe', they will have to be sufficiently 'European' themselves, i.e. be capable of thinking beyond the narrowly national perspective, so that they can act as a 'European partner' of Germany, not as its adversary (Handl *et al.*, 1997, pp. 170–1; Pick, 1997, pp. 13–14; Had *et al.*, 1997, p. 19).

The approach to Germany is characterized by a mix of geopolitical traditionalism, stressing Germany's size and proximity, and a more dynamic, 'institutionalist' perspective, stressing the new constellation in Europe with American presence, growing interdependence and integration, and shared democratic values (Eichler, 1997, pp. 67–8; Kotyk, 1997, p. 7; Pick, 1997, p. 11). Most authors operate with concepts such as 'Euro-Atlantic' or 'Euro-Asiatic' structures/civilization/framework, without discussing their status or validity. Winkler's not untypical formula varies the 'return to Europe' motif: 'the present situation in many ways provides an exceptional opportunity to anchor the Czech Republic in the Euro-Atlantic civilization where it belongs and where it wants to belong' (Winkler, 1996, p. 15).

Conversely, Russia is often defined as a 'Euro-Asian power' (Leška *et al.*, 1997, p. 88). When Pick boasts that the Czech Republic has decided to join the Euro-Atlantic space he adds that 'Slovakia has not succeeded in fully escaping from the habits and ties of its Euro-Asian dimension' (Pick, 1997, p. 14). He interprets political problems such as unstable democracy, incomplete privatization, etc. as a cultural problem: the influence of negative 'Euro-Asian qualities'.

The geopolitical perspective was particularly popular in the assessment of the consequences of partition from Slovakia. Many were jubilant, arguing with the government that, with the division, 'the Czech Republic moved geopolitically to the West' (Eichler, 1997, p. 67). It had 'detached itself from the economically and politically unstable, and potentially conflict-ridden, regions in Eastern and Southern Europe' (Winkler, 1996, p. 9; see also Pavlík, 1993, p. 21). Only Luňák challenged this view, when pointing out that 'it is no longer true that the degree of our approximation to the West is indirectly proportional to our distance from Russia' (Luňák, 1997, p. 198); instead, it would be in both Czech and Western interests for the country to have good relations with Russia.

The description of the Czech place in Europe was often cliché-ridden – as when the same Luňák declared that the Czech Republic had 'return[ed] to its natural cultural–political environment, incarnated in the Western European and Trans-Atlantic structures' (ibid., p. 185). Eichler displays a whole catalogue of conventional stereotypes, when he describes the Czech Republic as placed

> in the very centre of Europe . . . it finds itself at the borderline between two European cultures – the Western and the Eastern. . . . The Czechs mostly subscribed to Western Europe and in reality they also were a part of this culture. But the West did not always take the Czech Lands to be an integral part of the Western world.
>
> (Eichler, 1997, p. 67)

Here we meet again the traditional ambiguity as to whether the Czechs already are, or only aspire to become, part of the Western community. So, in the end, we may conclude that, even in the expert view, notions of civilizational roots precede any pragmatic political or economic considerations in the argument for EU integration.

Public opinions

Inevitably, public attitudes to the EU are very difficult to map. Judging by the coverage in the Czech media, the EU and its enlargement has not so far been at the centre of public debate, and although one can register broad support for the idea of Czech membership in opinion polls it seems fair to suppose that for the general public the issue is rather remote and abstract. The present discussion therefore restricts itself to some comments on what 'informed public opinion', as it appears in some select newspapers and journals, has had to say about this question.[8]

By far the most intensive coverage of EU affairs is found in the weekly *EKONOM*. EU news is well covered from a traditional, journalistic angle, in regular reporting from Brussels, and in interviews with Czech and foreign EU experts or political actors. The journal also offers more comprehensive background information on the EU. Thus in 1996 it published a series of articles on the history and functioning of the EC/EU. Starting in September 1997 it has published a bi-weekly supplement, *The Integration of the Czech Republic into the European Union*, in cooperation with the delegation of the EU Commission in Prague and supported from the Phare programme.

EKONOM thus openly supports the official agenda on enlargement as defined by the EU and the Czech government, although criticism of the EU certainly appears in specific cases: for instance Libuše Bautzová's article on how customs regulations dictated by the Europe Agreement have a negative effect on Czech exports. There the author concludes that the EU profits more from these agreements than the Central and Eastern European countries, *EKONOM*, 13/1997, pp. 22–3). Yet, in editorials as well as in articles and interviews, the general tenor of *EKONOM* is unconditionally 'pro-European'.

This also shows up at the terminological level, where in most cases 'Europe' simply refers to the EU: witness headlines like 'Czech Law is Heading Towards Europe' (*EKONOM*, 16/1996, p. 65), or 'Will Europe Swallow Us?' (*EKONOM* 31/1997, p. 20), a call for an increase in the productivity of Czech labour. Interviews and commentaries repeatedly describe membership as a 'great chance', economically and politically, for which the Czech Republic has to prepare. An example is an editorial from 1996, when there was still uncertainty about when enlargement talks would be commenced: Dušan Provazník claimed, in the 'no alternative' vein, that 'the main trend is given, so it is not so important when our approximation to the EU is sanctioned in a formal agreement' ('Our Europe', *EKONOM* 4/1996, p. 3). Two years later Eva Klvačová was very frank about the motives guiding the EU and the balance of power between the Union and the applicant countries:

> The entry of the Czech Republic into the European Union will only happen when we become attractive for this community, when the advantages for its members are bigger than the expenses related to us. Not a minute before. In the meanwhile we can think about the fact that we have never been, and will never be, the creators of the rules, but only the ones that have to

> accept them without objection. For somehow the alternative of applying for entry into another community does not really exist at present. It seems that this fact is dawning upon us rather slowly.
> ('Waiting outside the gate', *EKONOM*, 42/1998, p. 3)

There is certainly some truth in the last part of the statement – at least to judge from developments in attitudes expressed in the daily *Lidové noviny* (*LN*). Its journalistic coverage of the EU has mostly been informative, without any openly normative evaluations. But a clear trend can be detected from editorials and the newspaper's debating pages. Following the prevailing trend on the political scene, the *LN* has moved from a 'Klaus-like' self-confidence, openly critical of the Maastricht Treaty and the idea of the EMU, to an ever-greater humility and national self-criticism, bordering on self-flagellation.

Around 1995 the tone was set by the historian Dušan Třeštík and the philosopher Václav Bělohradský, two leading members of the newspaper's editorial board. Both saw the 'German problem' as the key not only to Czech foreign-policy interests, but also to the understanding of the EU and Europe in general. They shared a very critical attitude to Germany and the Kohl government, as in the following statement by Bělohradský, reacting to the allegation that Czech–German relations had cooled after Klaus's attacks on the EU:

> How quickly Chancellor Kohl has taken Brezhnev's place: Mr. Klaus has allowed himself to criticize not just the European Union, which Kohl considers to be his property, but even the ideology of Kohl's party! Mr. Chancellor obviously has other ideas about democracy than we do.
> ('Europe as a Threat', *LN*, 14 June 1995, p. 5)

To Bělohradský, real Europe incarnating the idea of the civic, heterogeneous nation-state, consists of 'the West' plus (giving priority to victors of World War II) Russia. Maastricht thus appears as one big, misleading attempt to solve the 'German question' by building 'in contradiction of a thousand-year-old tradition of European variety, the Maastricht monstrosity, the European super state'. Only if Germany completed its 'Westernization' and 'Europeanization' – by giving up its ethnic definition of the nation, its Central European ambitions, and its idea of Euro-regionalism – could the fears that led to Maastricht be removed (Bělohradský, 1996; see also his 'The

Owners of the Keys to the European Home?', *LN*, 18 May 1995, p. 8). Still, Bělohradský believed that the Czechs had 'deep civilizational reasons' for wishing to join the EU, but it remained their right and duty to criticize whatever, like the CAP, was immoral and wrong in the Union:

> If a critique of *status quo* really were a reason to refuse us membership of the EU, if the EU demanded an uncritical attitude from its members, then we would have no deeper civilizational reason for integrating into the EU. But I assure my readers that the EU does not fear critique.
>
> ('Fawning into Europe?', *LN*, 12 February 1996, p. 1)

Dušan Třeštík was equally sceptical of Maastricht, but he described it as a German idea (just like the 'Europe of regions' and a 'federal Europe'), popular only in a country that feared its own role as a nation-state, because it had so far been unable to emancipate itself from its typically *Gross*-German ethnic concept of citizenship ('Today Answers', *LN*, 12 December 1995). Třeštík felt confident that the Czechs could afford an independent stance, since Germany in any case had a 'vital interest . . . in stable states, firmly integrated into Europe, at its Eastern border' ('Chancellor Kohl's Brief Moral Episode', *LN*, 21 March 1995, p. 5).

Two years later new authors and attitudes characterized the journal. By the end of 1998 many articles on the EU were accompanied by the slogan '*Lidové noviny* supports the Czech effort to join the European Union'. Front-page editorials on the Commission's 1998 Progress Report had headlines such as 'The Vagabond and the Fifteen Giants', 'We are the Worst', or 'Czechs, Pull Yourselves Together' (*LN*, 21 October, 3 November, and 4 November 1998), while a background article was entitled 'The Czechs Outside the Union? We Would Be Like Turkey' (*LN*, 4 November 1998, p. 3). There was still room for dissent, as in the economist Karel Kříž's article 'Membership in the European Union is not a Fatal Thing' (*LN*, 10 November 1998, p. 11), but the article immediately above it by Michal Musil, assistant editor-in-chief, was more typical. Musil strongly attacked the Czech Eurosceptics, with Klaus at the head, for distorting the facts about the Commission Report and the Czech position: the EU was fully entitled to tell the Czechs what to do to live up to the rules that had functioned for decades in the community that they wanted to join. These rules were binding, and if the

Eurosceptics did not want to adhere to them, at least they could openly declare their 'no' to membership ('Our Smart and Lazy Eurosceptics', ibid.).

Bohumil Doležal, the journal's new chief commentator, and a sharp critic (in other newspapers) of the Bělohradský/Třeštík/Klaus foreign-policy line (Doležal, 1997), was very pessimistic. Western European politicians had some responsibility for creating unrealistic hopes in the Central and Eastern European countries, but Doležal put the main blame on 'irresponsible politicians' such as Klaus and Havel, who had nourished the illusion that the Czechs already were European, so that the nation was blind to the huge work that had to be done to reduce the enormous distance that separated all post-communist countries from the EU: 'We all have to get used to the view that our integration into the EU belongs to a very, very distant future, and that perhaps we will never get into the EU' ('Our Membership of the European Union has become a Vision', *LN*, 30 November 1998, p. 10).

Such huge fluctuations of opinion have not occurred in the weekly *Respekt*. Its staff has changed less, but it is probably more important that the journal soon adopted a 'realistic' (i.e. pessimistic) view of the prospects of Czech membership of the EU, and a coherent idea of what the Czechs should do regardless. *Respekt*'s coverage of EU matters has not been as intensive as in *EKONOM*, but there has been regular reporting and room for larger analytical background articles, including serious discussions of motives and actors in the EU influencing decision-making on enlargement.

The journal's approach was sketched out in Zbyněk Petráček's article 'Adventures in No-Man's Land', from 1994. Petráček argued that the 'road to Europe', as defined by the EU, would be long and hard. So, instead of calling for an illusory early entry, the Czechs should think how to act in their 'grey zone' between the EU and Russia. Membership ought still to be a long-term goal, for economic, political and security reasons, and because of the wish to become finally and definitely a 'normal state'. But the EU had many good reasons for not being in a hurry, considering the present abyss between the EU and the post-communist countries, so the Czechs had to be patient (itself 'no disaster'), while doing their utmost to get as close as possible to Western legislation, economy, administration, and cultural behaviour (*Respekt*, 36/1994, pp. 7–9).

Two weeks later Jan Macháček analysed the consequences of a hypothetical early acceptance, and listed a series of risks for Czech agriculture and industry, including the risk of a 'brain drain' ('How Would We Do in the European Union?', *Respekt*, 38/1994, pp. 2–3). Macháček returned to this theme again in 1996, as the Czechs applied for membership, with a lengthy feature. He discussed the many internal differences, institutions and interests within the EU that militated against an early enlargement, and listed the many sectors where the Czechs were still unprepared for membership – including such cultural factors as a pronounced xenophobia and fear of the Germans. Still, the author tried to strike an optimistic note:

> [How shall we] answer the question of what to do if we really do not get into the EU in the coming decades. It is simple: we can express our affinity to the West one way or another. Even without membership of the Union we can let EU citizens buy real estate on our territory. We can further liberalize the convertibility of the Crown, reduce taxation and simplify legislation just the same. Just as nothing prevents us from equipping our toilets with paper and stopping the shameful robbery of 'wealthy tourists' with double prices. Every major and minor foreign investment also brings us closer to the West, because it creates an economic and a closely linked political interest. . . . If we want to, we can become a self-conscious part of Europe even without the trademark membership of their elite club.
>
> ('Good Morning, Europe', *Respekt*, 4/1996, pp. 9–11, quotation from p. 11)

The journal was far from uncritical in its comments, and the EMU was long considered to be a hazardous and very uncertain project. But as Petráček remarked, even if one found that the Maastricht criteria were a strange, if not socialist, piece of engineering, the fact remained that fifteen European states had accepted them, and so there was nothing to do but to behave as if they were binding also for the Czech Republic ('Don't Speculate, Act', *Respekt*, 26/1997, p. 2). So the Commission's 1998 Report came as no shock to Macháček, who found that the Czechs had quickly to put right what was being criticized, 'whether or not we want to join the Union – simply because we want to live in a civilized country' ('The Gospel from Brussels', *Respekt*, 46/1998, p. 2). From *Respekt*'s point of view there might be an alternative to EU membership (including exclusion from the EU), but not to a 'Westernization' which by and large equalled the copying of the EU paragon in any case.

Conclusion

Looking back, the evolving views among politicians and the public alike seem almost paradoxical: the closer the Czech Republic has come to membership, the more the picture of the Czech position *vis-à-vis* the EU has been marked by frustration and pessimism. The 'equal partners' perspective seems totally gone.

This may be ascribed to domestic political developments. The euphoria of 1989, and the self-awareness that followed the Czech 'liberation' from the Slovak 'East', have been replaced by political instability and economic stagnation. Jiří Pehe has suggested that the present gloom may be seen as a product of the efficiency with which Czech politicians and diplomats after 1993 presented an idealized picture of Czech realities, based on the 'ideology of uniqueness and success, cultivated by the Klaus government'. As the Czech Republic proved unable to live up to these high expectations, a 'swing of the pendulum' occurred in foreign and domestic perceptions alike (Pehe, 1998, pp. 61–2). Today, Poland and Hungary have taken over the role as Central European favourites, which again allows their governments to criticize the EU, while the Czechs humbly accept all verdicts coming from Brussels.

'European Visions have been replaced by Emptiness', the *LN* claimed, as it compared earlier proclamations of Havel and Klaus to contemporary statements (*LN*, 4 November 1998, p. 3). But after the initial, 'utopian' post-1989 optimism confronted reality in ten years' experience with the EC/EU, this hardly comes as a surprise. Gradually it has dawned on the Czechs that many a grand declaration from the West has covered a reality marked by protectionism, and a very low commitment to Eastern enlargement. It took the EU four years even to recognize the *idea* of enlargement, and eight years before concrete negotiations could begin, accompanied by demands so tough that they border on the impossible. 'Europe' has not ceased to be divided, and for all their claims to the opposite the Czechs are still on the wrong side of the dividing line.

For this reason many Czech statements have been characterized by a limited knowledge of what the EU was actually like. At times the EU has functioned more as a mirror for the national self-perception than as a complex, concrete entity on the foreign political scene. So 'Maastricht' has often been used as a metaphor or symbol of a 'socialist', 'technocratic', 'unnatural' threat to the nation-state and the free market that the Czechs have come to love after their

'Soviet' experience. Or, alternatively, the EU is glorified as the saviour of the Eastern Europeans from their own demons. The lack of detailed knowledge about the EU, and the feeling of insecurity, have also made the discussion very vague. The typical 'yes' to the EU is explained with generalized 'national interests', while an in-depth discussion of the positive and negative effects of membership on various sectors of Czech economy and society is largely absent.

For all the occasional verbal 'strongmanship', national self-confidence has not prevailed in the debate, neither politically nor in terms of self-identity. The Czech attempts to derive a right to membership from history have suffered from the fact that the West often perceived things quite differently, so that claims to be thoroughly European in the past look like a poor attempt to make up for shortcomings of today. Hence, the constant fear of being excluded, visible even in the self-assured rhetoric of Václav Klaus. As such, the Czechs seem acutely aware of the discrepancy between the claimed 'belonging to Europe' at the ideal level, and the reality of their peripheral position today.

Notes

1 To a large extent this chapter is based on research conducted during a three-month stay at the Robert Schuman Centre of the European University Institute in Florence in early 1998. I am grateful to the EUI for the research grant given on that occasion. Many of the ideas expressed here have also been discussed in a Robert Schuman Centre Working Paper resulting from this stay, 'Czech Perceptions of the Perspective of EU Membership – Havel vs. Klaus' (1999).

2 This new, positive view of the present EU may reflect the voice of Foreign Minister Josef Zieleniec more than a change of attitude on the part of Václav Klaus. According to Jiří Pehe, Czech foreign policy in these years had three main sources: the Prime Minister, the President, and the Minister of Foreign Affairs, the latter two differing from Klaus's approach on the EU and on Czech–German matters (Pehe, 1998, p. 63; see also Had *et al.*, 1997, p. 52). Thus Zieleniec praised the political dimension of EU cooperation, especially its importance for preserving peace and preventing 'national wars' in Europe. Nor was he afraid of any development towards a 'United States of Europe' run by a supranational bureaucracy.

3 Ironically, especially since the mid-1980s, the EC had invested a lot in this kind of 'identity politics', similar to nation-building – complete with flag, anthem and history books – but with very poor results. See Boxhoorn (1996) and Smith (1992).

4 Klaus was much in favour of a 'multi-speed Europe', since it meant giving up the ambition to create one binding model for all and thus facilitated Czech integration into the EU. 'Widening' rather than 'deepening' was

his priority, because, as he put it, it was easier to get on a slow train than on an express train (Klaus, p. 138). Thus his resistance to the Maastricht Treaty seems to stem also from a fear that the concentration on internal affairs may make the EU forget Eastern Europe (Klaus, 1994, p. 148).

5 After the 1996 Czech application for membership Klaus became less militant in his attacks on the EU, but his basic opinions remained the same. He protested against the idea that the inclusion of new members in itself justified more majority voting, and he was sceptical of the consequences of the EMU. But at the same time he expressed his fear that the EU would prefer a 'deepening' of integration before enlargement, without even openly discussing its priorities or the interdependence of the two processes ('This Week Was Full of European Things,' *Lidové noviny*, 27 January 1996, p. 8; Klaus, 1997, pp. 1–2).

6 I shall not discuss Robejšek's claims about the academic standards of Czech foreign policy expertise. But one must agree with the author that the expert environment is quite narrow. This is due to the near-complete ban on political science before 1989, and also to the rather low priority given to higher education in general and political science and European or international studies in particular, even after 1989. In recent years initiatives have been made, but it takes time to educate a new academic elite. Furthermore, many of the contributors to the Czech foreign-policy debate are directly or indirectly affiliated with the Ministry of Foreign Affairs, which cannot but affect their perception and attitude.

7 Others follow the 'no-alternative' line of the government, without much explanation. Without elaborating on why, Jakš calls membership 'objectively a clear *sine qua non* for the long-term prospects of the transformation process in the Czech Republic' (Jakš, 1994, p. 143). Urban insists, given the Czech Republic's geopolitical position, economic level, and degree of integration into the international economy, 'one can hardly imagine any rational alternative to this step' (Urban, 1995, p. 9).

8 My coverage of the Czech media landscape is necessarily incomplete, and the selection of sources to some extent random, determined by what I have had access to. The debate has been followed in *Lidové noviny*, a daily which, with its background in the pre-1989 dissident milieu, strives to have an intellectual profile; in the political-cultural weekly *Respekt*, which addresses a similar, mostly well-educated readership; and finally in the weekly *EKONOM*, which – as suggested by its name – specializes in economic news. All three journals have a liberal/ring-wing orientation, a bias from which the survey may also suffer.

References

Agenda 2000 – Commission Opinion on the Czech Republic's Application for Membership of the European Union, DOC/97/17, Brussels, 15 July 1997.

Bělohradský, Václav (1996) 'Proti státu Evropa', *Literární noviny*, 51–52/1996, 1 and 3–4.

Boxhoorn, Bram (1996) 'European Identity and the Process of European Unification: Compatible Notions?', in M. Wintle (ed.), *Culture and Identity*

in Europe (Avebury, Aldershot), pp. 133–45.
Břach, Radko (1992) *Die Außenpolitik der Tschechoslowakei zur Zeit der 'Regierung der nationalen Verständigung'*. Schriftenreihe des Bundesinstituts für ostwissenschaftliche und internationale Studien, Köln, vol. 22 (Baden-Baden: Nomos).
Desný, Petr (1997) 'The Harmonization of the Legislation of the Czech Republic with European Union Law', *Perspectives*, 8 (Summer), 45–54.
Doležal, Bohumil (1997) *Nesamozejmá politika* (Prague: Torst).
Eichler, Jan (1997) 'Priority zajistení bezpečnosti České republiky', in V. Kotyk (ed.), *Česka zahraniční politika: Uvahy o prioritách* (Prague: Ustav mezinarodních vztahů), pp. 61–85.
EKONOM, 1996–98.
Policy Statement of the Government of the Czech Republic, 27 January 1998, from: http://www.vlada.cz/vlada/dokumenty/progrproh.eng.htm (as of February 1998).
Policy Statement of the Government of the Czech Republic, August 1998, from: http://www.vlada.cz/vlada/dokumenty/prohlas.eng.htm (as of December 1998).
Had, Miloslav *et al.* (1997) 'Česká republika a evropská integrace', in Kotyk (ed.), *Česka zahraniční politika: Uvahy o prioritách* (Prague: Ustav mezinarodních vztahů), pp. 17–60.
Handl, Vladimír (1995) 'Translating the Czech Vision of Europe into Foreign Policy – Historical Conditions and Current Approaches', in B. Lippert and H. Schneider (eds), *Monitoring Association and Beyond: the European Union and the Visegrád States*, Europäische Schriften des Instituts für Europäische Politik, vol. 74 (Bonn: Europa Union Verlag).
Handl, Vladimír *et al.* (1997) 'Česká republika a Německo', in V. Kotyk (ed.), pp. 153–83.
Havel, Václav (1991) *Letní přemítání* (Prague: Odeon).
Havel, Václav (1992) *Vážení obcané (projevy červenec 1990 – cervenec 1992)* (Prague: Lidové noviny).
Havel, Václav (1995) *'94* (Prague/Litomyšl: Paseka).
Havel, Václav (1996) 'Projev na slavnostním zahájení kongresu Nové atlantické iniciativy 10.5.1996', *Střední Evropa*, 61/1996, 77–80.
Havel, Václav (1997) *'96* (Prague/Litomyšl: Paseka).
Jakš, Jaroslav (1994) 'The Czech Republic on the Road to the European Union – Problems of the Mutual Interaction of the Transformation and Integration Processes in the 1990s', *Perspectives*, 3 (Summer), 141–9.
Klaus, Václav (1994) *Česká cesta* (Prague: Profile).
Klaus, Václav (1995) *Dopočitávání do jedné* (Prague: Management Press).
Klaus, Václav (1997) 'Europe on the Homestretch'. Keynote speech given at the 7th Frankfurt European Banking Congress, 27 November, from: http://www.vlada.cz/historie/vlada97/projevy/frankfurt.eng.htm (as of February 1998).
Kotyk, Václav (1995) 'K problematice zahraničněpolitické koncepce Ceské republiky', *Mezinárodní vztahy*, **4**, 65–73.
Kotyk, Václav (ed.) (1997) *Česká zahraniční politika: Úvahy o prioritách* (Prague: Ústav mezinárodních vztahů).
Kotyk, Václav (1998) 'Courage in Political Science and Czech Foreign Policy', *Perspectives*, **10** (Summer), 57–60.

Leška, Vladimír (1996) 'The Future of Czech–Slovak Relations in the Light of European Integration', *Perspectives*, **6–7**, 75–87.
Leška, Vladimír *et al.* (1997) 'Česká republika a region střední Evropy', in V. Kotyk (ed.), *Česká zahraniční politika: Úvahy o prioritách* (Prague: Ústav mezinárodních vztahů), pp. 87–124.
Lidové noviny, 1994–98.
Luňák, Petr (1997) 'Postsovětsky prostor v prioritách české zahraniční politiky', in V. Kotyk (ed.), *Česká zahraniční polítika: Úvahy o prioritách* (Prague: Ústav mezinarodních vztahů), pp. 185–203.
Memorandum vlády České republiky k žádosti o přijetí do Evropské unie (23 January 1996), in V. Kotyk (ed.), *Česká zahraniční politika: Úvahy o prioritách* (Prague: Ústav mezinárodních vztahů), pp. 259–60.
Moorhouse, Jacqui (1996) 'Osteuropa auf dem Weg in die Europäische Union', *Aussenpolitik*, **IV**, 367–78.
Pavlík, Petr (1993) 'Začleňování ČR do evropských a světových hospodárských struktur', *Mezinárodní vztahy*, **2**, 65–73.
Pech, Radek and Winkler, Jan (1996) 'Česká zahraniční politika v období 1993–1996', *Mezinárodní politika*, **4**, 4–7.
Pehe, Jiří (1998) 'Connections Between Domestic and Foreign Policy', *Perspectives*, **10** (Summer), 61–4.
Pick, Otto (1997) 'Česká republika ve světě', in V. Kotyk (ed.), *Česká zahraniční politika: Úvahy o prioritách* (Prague: Ústav mezinărodních vztahů), pp. 11–16.
Respekt, 1994–98.
Rhodes, Matthew (1999) 'Post-Visegrad Cooperation in East Central Europe', *East European Quarterly*, XXXIII, no. 1, pp. 51–67.
Robejšek, Petr (1998) 'The Imperative that Foreign Policy Must Depend on Internal Policy', *Perspectives*, **10** (Summer), 75–9.
Šedivý, Jiří (1994/95) 'From Dreaming to Realism – Czechoslovak Security Policy', *Perspectives*, **4** (Winter), 61–71.
Šedivý, Jaroslav (1997) *Černinsky palác v roce nula (ze zákulisí polistopadové politiky)* (Prague: Ivo Železny).
Šedivý, Jaroslav (1998) 'The State of Czech Foreign Policy and its Prospects for 1998', *Perspectives*, **10** (Summer), 5–10.
Smith, Anthony D. (1992) 'National Identity and the Idea of European Unity', *International Affairs*, **28**(1), 55–76.
Urban, Luděk (1995) 'Bílá kniha – cesta přidruženych zemí do vnitřního trhu Evropské unie', *Mezinárodní vztahy*, **4**, 5–11.
Winkler, Jan (1996) 'From the Partitions to the Elections: the First Years of Czech Foreign Policy', *Perspectives*, **6–7**, 7–15.
Zemánek, Jiří *et al.* (1997) 'Status and Tendencies in the Czech Republic', in P.-C. Müller-Graff (ed.), *East Central Europe and the European Union: from Europe Agreements to a Member Status*, ECSA – Series, vol. 5 (Baden-Baden: Nomos).
Zieleniec, Joséf (1993) 'Ke koncepci zahraniční politiky České republiky, 21.4.1993', in V. Kotyk (ed.), *Česká zahraniční polítika: Úvahy o prioritách* (Prague: Ústav mezinárodních vztahů), pp. 243–50.

11
Historical Memory and the Boundaries of European Integration

Hugo Frey

> In May 1945, three weeks after the capitulation, a German university teacher of history is said to have declared: 'There is no sense in teaching German history any more. There is no longer any German history anymore.' Such a statement sounds strange indeed in English ears, for history has never been with us, at any rate consciously, an instrument of state policy or the handmaid of a particular *Weltanschauung* or political viewpoint.
>
> (Bing, 1951, p. 92)

The above statement on the political uses of history opened Harold F. Bing's 1951 assessment of the study and teaching of the subject in post-war Germany. Published in an eminent journal of the day, Bing's article provides a snapshot of the British profession's perception of contemporary German scholarship. Moreover, it contains a detailed overview of Germany's tentative first steps towards offering an analysis of the Third Reich. Reporting on an international academic conference devoted to these subjects, the article is a reminder of the uncertainties of West European reconstruction: intellectual and practical. For instance, Bing posed the questions: how would history be taught to a defeated nation; what efforts were the German academic elite making to provide a new civic culture? Equally, how were British historians to help in the re-creation of German education? For example, they could provide instruction and guidance in forums such as the conference, which took place in the 'British zone'.

Bing's essay reveals the gulf in the appreciation of history that existed between the 'winners' and 'losers' of World War II. On the one hand a devastated, divided and occupied Germany considered what demoniac forces had led to the disasters of 1945. The nation's historians debated which philosophy of history would best serve the post-war world. Rightly or wrongly they worried that Nazism had compromised their profession. Nonetheless, they also wished broadly to support new routes to social and educational normality. Defeat had thrown the world of the German historian into turmoil. War had destroyed schools, universities and other buildings and even basic textbooks were scarce. More importantly for the academic discipline, the course of recent history had yet to be adequately explained. The teachers in particular feared that what they had learnt during the 1930s was no longer acceptable or accurate. As Bing reported to his readership, the history teacher 'could teach only what he had been taught, and so far as the youngest teachers were concerned what they had been taught was politically biased and was now officially disapproved' (p. 104).

Observers from the outside, like Bing, a representative of the Anglo-Saxon view, probed and analysed the German problem in all its forms. As Pierre Ayçoberry and others have noted, there was a tendency in the early non-German literature on Nazism to portray the Reich as a stricken patient that was awaiting therapy (Ayçoberry, 1981). The tone adopted by Bing in his article was no exception. In startlingly condescending prose he suggested that a fundamental problem that his German colleagues faced was the inherent authoritarianism of their educational system. To paraphrase, he distinguished the fact that the educators feared to admit that sometimes they did not know everything about their given specialism. The implication was that in such a system any recognition of weakness would begin to undermine the power relations that existed between teachers and students. According to Bing, the 'assumed omniscience' of the German professor had to be adapted to accept an implicitly more Anglo-Saxon willingness to admit ignorance. More importantly, the British writer suggested that the Germans also had to realize that the historical world went far beyond the rivers Rhine and the Oder. Praising the curricula which were taught in primary schools from London to Edinburgh, he explained that 'we do not limit ourselves to the locality. We have also our "Stories from World History" introducing the heroes of all the nations, and in the geography lessons we describe the children of other lands, so that our pupils learn something of the lives of Esquimaux, or Chinese or African children' (Bing, 1951, p. 106). Of course, the commentary says at least as much about the character of British academia in the 1950s as it illustrates the perceived limitations of instruction in Germany with its focus on the *Heimat*.

The *petite-histoire* of 1950s intra-European historiographic exchanges may appear rather distant from topics such as centre–periphery dynamics, or the continent's margins and borders, which are discussed elsewhere in this collection. Nonetheless, I suggest that its relevance lies as a pointed illustration of how, while European integration was developing in the economic field, it has not developed in the same way in other sectors of sociocultural life. Even as Monnet, Pleven and the other early federalists were setting up the Coal and Steel Community, cultural and intellectual harmonization evoked a complex interplay of identities and discourses between *self-differentiating* Europeans.

This chapter addresses how memories of World War II have both supported and worked against the development of West European integration or any common conception of what is Europe. The terms 'memory' and 'historical memory' are used interchangeably to mean the manner in which the past is socially interpreted: that is to say how civic and political groups retrospectively select and view important events symbolically. Such 'memories' are communicated via all manner of cultural products, including programmatic political statements, popular journalism, film, literature, high art, as well as historiography itself. Ultimately, they must contribute to the self-identification of Europeans, together or as separate groups, and hence also to the 'geography' of historic borders attributed either to Europeans in common or at odds with each other.

First, I examine the recent interpretations offered by Tony Judt (1997). Judt's thinking is just a widely read comment on the presupposition that post-war European unification must have been backed by some kind of cultural unity – specifically, a common treatment of Europe's recent historical past. Judt offers an up-to-date, brilliantly argued exposé and critique of the view that memories of fascism and the war period might still promote a collective desire to unite. Evaluating this outlook, the rest of the chapter reflects on the growing specialist literature on historical memory, which suggests a quite different viewpoint. I argue that examples of memories of the war, like the Bing article, undermine the *initial* expectation in Judt and others: that a shared West European identity or common space could be based in the first place on concordant historical memories.

Europe and history: the Judt thesis

Like Bing in 1951, sometimes historians consider that they should speak out and address the wider public whom they feel that they serve. This is especially the case when they are tempted to comment on contemporary political trends and to place them in their wider context. One good example of such an intervention is Tony Judt's *A Grand Illusion? An Essay on Europe* (1997). Published in a popular paperback format, Judt's book offers a 150-page review of contemporary European affairs. This is an eminent historian's contribution to the issues of the day. Judt divides the text into four commentaries that, although dealing with a variety of topics, provide a coherent view of the continent and its prospects.

Chapter 1, which also bears the title of the collection, is an informed presentation of the origins of post-war integration and the creation of the European Communities. Drawing extensively on the interpretation already established in the research of Alan S. Milward (1984, 1992), Judt tells the reader that nation-state self-interests, rather than idealistic federalism, drove the creation of the EEC. Reviewing the context in which the nations were prepared to pool resources and governmental operations, he argues that the period between 1945 and 1989 was especially conducive to institutional collaboration in Western Europe. Taking account of specific policy areas, notably the CAP, and more general trends, it is suggested that the environment in which the Treaty of Rome was signed and the EC flourished was remarkable and unique. Factors as diverse as the need for economic reconstruction, the international relations conjuncture and historical memory of the war all worked in favour of cooperation. However, Judt infers that all these conditions for further unity via EU structures ended with the turbulent collapse of Communism and the return of Central and Eastern Europe to the international arena. Judt remarks, 'the illusion of Europe would not survive being put to a continent-wide test' (p. 44). The pressures of a post-communist Central European enlargement would be too great for the Western club to continue as if nothing had changed. The period of relative stability, affluence and comfort for the West Europeans will be exchanged for a more complex future.

The second chapter takes us to the issue of borders: the symbolic location of Europe and where one may fix its many possible boundaries. Judt treats the eternal question: where does Europe begin and end? Unsurprisingly, given the attitudes expressed in the previous chapter, he takes an expansive geographical definition that looks beyond the architecture of Brussels and Strasbourg. But within that space he stresses that Central and Eastern Europe have distinctive heritages and traditions that need to be recognized by those of us living in the West. Repeating the pessimistic conclusions of the previous essay, he suggests that EU enlargement will carry over the long-standing prejudices through which the eastern edge of the continent has frequently been viewed. He argues that the post-communist societies 'in transition' would not be well advised to abandon their newfound national freedoms in exchange for the gloss of EU membership. Only inclusion in the institutions on an equal footing with their richer neighbours would merit such a decision, and this appears unlikely given the economic inequalities and prejudices that would have to be bridged in this scenario. Western Europe will continue to construct its own borders around the supposed backwardness of the East.

'Goodbye to All That?' is the final major chapter of the collection. Combining several themes which were only implicitly raised in the earlier sections of the book, Judt argues that '1989' represents a major threshold in European history. Not so much the end of History, as the concluding point of the post-war historical moment. Since the reunification of Germany, the fall of the Soviet Empire and the shift from a bipolar to a unipolar world, international politics and internal European society will never be the same again. To paraphrase, the West Europeans have now to realize that those cultural, political and international factors that once gave birth to the EU will soon fade. Abandoning the cautious tone of argument displayed in his handling of the East/West question in his Chapter 2, Judt claims that the nation-state will soon be making a robust return to the stage. He argues that middle-sized national governmental units can serve their citizens in ways which are more adept than transnational or regional structures. Whilst I can neither prove nor dispel the prediction, the type of historical thinking behind Judt's preferences as to form of government, and his sceptical attitude towards the European Union and its future prospects, are revealing. This is how he explains his thinking:

> It may well be that the nation – with the *community of memory* that it represents and the state that embodies it, with its familiar and appropriately scaled frame – is the only remaining, as well as best adapted, source of collective and communal identification . . .

> the years after 1989 will require a *rehabilitation* of the nation-state's political and cultural credibility if Europe is to remain afloat.
> (pp. 119–21; emphasis added)

I will take up this mode of thought below.

Judt supports his prognosis by showing that Europe (in the guise of the EU and the Schengen group) is no more pluralistic and welcoming to outsiders than were the exclusive nation-states of the past. Sharply, he taxes those who have claimed that the EU represents a new, more tolerant society with its stance on immigration. In a tone that has echoes of British Eurosceptic literature (see Christopher Flood's chapter in this book), he remembers that the term 'Fortress Europe', which emerged in the Schengen debates, was originally coined by Hitler's propaganda minister, Joseph Goebbels. Of course, there is no intention to suggest that West European integration is a German conspiracy or a continuation of a Nazi project. Judt is making fun of those Europhiles who only ever interpret the EU's actions through rose-tinted spectacles.[1] European integration, he implies, has produced the same historic intolerance to outsiders, without the nation-state's saving grace of internal cohesion.

An impressive intellectual performance in its own right, *A Grand Illusion?* is the work of a self-professed 'Euro-pessimist' (p. 1) who nevertheless admires the work of the Union. In the western quadrant of Europe there has been relative economic success, social stability, peace and security. These achievements are politely applauded. The book, like many plausible historical interpretations, is a middle-way account that sees integration as a process to be treated neither with 'bland assurance' nor with 'dire prophecy' (p. 129). In the place of these two extremes several major themes run across Judt's argument: (a) West European integration was achieved only because of a quite particular set of historical circumstances; (b) that the events of 1989 shattered that context, so that the current climate is more conducive to a retrenching of the EU and a revival of nation-states; and (c) that, given long-standing historical conditions, the boundaries and status of its eastern borderlands remain a challenge to any unification of Europe.

Historical memory in Judt's argument

Judt's discussion of the role of historical memory is critical to each of the above hypotheses and supports his analysis of integration *per se*. The legacies of the past, treated across the essays and mentioned in no fewer than ten different contexts, are indeed central to the interpretation of European politics on offer. By talking so frequently of collective memory, memory or amnesia, Judt himself suggests that these issues are decisive. War-memory is ascribed a key role in accounting for the period of EU development (1945–89) and in the prognosis offered on the post-1989 world. Equally, memory is raised as an important influencing factor in Judt's discussion of other issues, e.g. the formation of the Welfare State consensus (p. 95) or contemporary German foreign policy (p. 137). Finally, in the quotation cited above, it is argued that the suggested revival of the nation-state is in part based on its ability to establish a 'community of memory'.

Specifically, Judt claims that a 'West European' collective memory of World War II was critical to the launch, development and success of the EU. A shared forgetting of the 1939–45 conflict was equally important as the push for economic reconstruction or the external pressures of the Cold War in stimulating integration. Through a common sense of shame towards either fascism in power, or of occupation and collaboration, the Europeans collectively dispelled the memory of the war. This *Zeitgeist* underwrote a context for unity and Franco-German cooperation. As Judt pithily writes, 'Hitler's lasting gift to Europe was thus the degree to which he and his collaborators made it impossible henceforth to dwell with comfort upon the past' (p. 27, see also pp. 37 and 83). He suggests that Europeans did not wish to recollect the traumas of contemporary history. More concerned with economic and social modernization, the French, the Germans, the Dutch, the Belgians and the Italians overlooked the horrors of the wartime period. This is an adaptation of the ground-breaking work of Henry Rousso on the collective memory of Vichy France (1990), which has to an extent already been repeated by Alan Morris (1992). In Judt's version only with the acceptance to leave 'skeletons in the cupboard' could the Europeans bring themselves to work more closely together. Through a shared aversion towards their recent history, West Europeans established a pact of silence: the values of progress in the present were allowed to oust memories, and guilts from the past. To prove his case dramatically, Judt cites a controversial 1969 speech from the leader of the Bavarian Christian Social Union, Franz Josef Strauss, who remarked that 'a people who have achieved such economic performances have the right not to have to hear any more about Auschwitz'

(p. 36). Silenced memories were needed, it seems, to propagate European unity.

Judt contends that the post-1989 period contours of memory have shifted. In the 1990s scholarly, cultural and popular interest in the far right and nationalism has increased. Germany and France have both witnessed significant historiographic debates in which war issues that had been previously considered taboo were fully aired. Citing the work of Rousso directly, Judt suggests that a new consensus, willing to scrutinize Nazi-related history in more detail, has been firmly implanted. Whereas for 'four decades' (1945–85) Europeans felt uneasy about confronting topics such as Nazism, popular and intellectual regard for these subjects is now common. Judt finds a critical role of historical memory in the perceived refurbishment of the nation in the post-1989 European scene. He writes:

> Since 1989 there has been a return of memory and with it, and benefiting from it, a revival of the national units that framed and shaped that memory and give meaning to the collective past. This process threatens to undermine and substitute for the inadequacies of the Europe-without-a-past. Thus for many years, in France or Germany, nationalist rhetoric was discredited by its close association with the memory and language of Nazism or of the Pétain regime. . . . This self-censure has all but disappeared except among an older generation of left-wing intellectuals, nowadays largely ignored. After two decades during which identification with Europe seemed to be replacing association with a nation, 'Euro-barometer' opinion polls are suggesting a reverse trend. In Germany, Denmark, Spain, Portugal and the UK a majority or near majority of those asked in 1994 saw themselves in the coming years as identifying uniquely with their own nation.
>
> (p. 118)

So, for Judt, increased 'remembrance' of fascism, Petainism and Nazism has resulted in a revival of support for national communities. The West European nations had to face their shame before they could once again perceive themselves through more positive patriotic imagery.

A European historical memory?

We can see that, even beyond the confines of Judt's book, war memory is implicitly used to identify 'Europe'. The post-war understanding of 1939–45 is one strand of civic life which denotes Europeanness. Whether 'forgetting' or 'remembering', we might expect that Europeans would be culturally identifiable through any attitudes they share towards history. This collective past is potentially an important marker of European identity. The thinking behind this approach suggests that one might locate 'Europe', and that within it there have been common reactions to a heritage of violent conflict. Before returning to the broadly geographical questions about the nature of defining the border of Europe through historical memory, let us examine in more detail this contrast between amnesia and remembrance.

In Judt, historical memory is shown to have either supported integration (1945–89) or weakened it (1989 onwards). But I will argue that research revises crucial aspects of the argument. First, one finds qualitatively significant political phenomena that do not correspond to the chronological outline established in *A Grand Illusion?* Secondly, I will show that there is little evidence to suggest that contemporary memories of fascism, Nazism or collaboration have stimulated a revival of nationalism in the 1990s. The bulk of products which portray war history remain ardently anti-fascist in their tone. Conversely, works which in some ways rehabilitate fascism have had at least a limited presence within the European cultural scene long before the fall of the Berlin wall.

One very notable exception to the Judt approach is the Gaullists. This major ideological current in West European post-war politics did not at all chime with Judt's thesis. De Gaulle and his followers, on the stage from the liberation to the late 1960s, cannot be described as figures who lived in a 'Europe-without-a-past'. On the contrary, Gaullist politics in France was saturated with references, allusions and memories of the occupation period (see among others Azéma and Bédarida, 1994; Morris, 1992; Frey, 1999). Instead of turning their backs on the events of 1939–45 this influential group frequently used the heritage of the resistance to its advantage: notably during the notorious European Defence Community debates (1954) when Gaullists and French Communists opposed military integration with Germany. As has been suggested by Jean-Pierre Rioux, memories of the resistance meant that those two ideologies were not prepared to allow French military grandeur to be subsumed within a European framework (1984, 1987). Among other factors, the memories of heroic resistance meant that integration in this sector could not develop at an equivalent pace to either the ECSC or the subsequent EEC. De Gaulle favoured European integration, but on his own terms: limited and forged through his wartime defence of French independence.

Likewise, de Gaulle's later stance during the Luxembourg Crisis (1965) shows another robust defence of sovereignty. As is well known, the General refused to accept an increase in the supranational powers of the Communities, even though indirectly French farmers would have materially gained in what was an election year at home. Leaving the infamous 'empty chair', the French delegation temporarily withdrew from the Council of Ministers. This dramatic and highly symbolic act evoked the Gaullists' tradition of protecting national honour and impacted on integration for years after. Although not directly related to the heritage of the war, the events of autumn 1965 were in line with the General's personal sense of historic destiny. The duty of resistance, declared for the first time in June 1940, meant that France should remain independent, a sovereign state which would neither fall under the complete control of Nazism, nor subsequently Washington, Moscow or the putative supranationalism of Brussels. In short, as Harold Macmillan once dryly remarked of de Gaulle, 'he talks of Europe and means France' (cited by Lacouture, 1992, p. 345).

These examples show a scenario that does not register in Judt's analysis: positive memories of anti-fascist resistance that acted in support of nationalism. Across Europe the nation-state was not always systematically devalued. While the Holocaust discredited fascist nationalism, the French, the Italian, the Dutch and the Belgium resistance movements often saw themselves as the repositories of national honour. Sometimes, as in the famous case of Spinelli, this even led to a position which favoured a federal re-structuring of Europe. In other instances, notably that of de Gaulle, it meant a forceful defence of the nation alongside involvement in integration. Judt suggests that all the Europeans felt as if they 'had lost the war' or had been compromised by collaboration (p. 26). But de Gaulle and others did not interpret the past in the negative manner that Judt ascribes. Instead, the anti-fascist struggle provided a new framework for post-war nationalism, untainted by either collaboration or the Nazi project. Here, the memory of conflict did not act as stimulus for Europeanism but instead it warranted a vigorous patriotism, which could be pro- or anti-Europe depending on contingent circumstances. In short, *pace* Judt, one can note that memories of 1939–45 never collectively discredited nation-states in the period suggested.

Now let us go back to the 'return of memory' question. Certainly, an increased interest and awareness of issues related to Nazism and the Holocaust is an established trend in the 1990s. Several European states have witnessed soul-searching re-evaluations of their wartime record. This was already 'in the air' at the time of the Bitburg controversy of 1985: when President Reagan visited a German military cemetery that included the graves of SS officers. The German 'Historians' Debate' of 1986 – notably in the press exchanges between Nolte and Habermas – again brought historical memory of Nazism to the centre of political and intellectual life (see Bosworth, 1992; Hartmann, 1986; Maier, 1988). While France has not seen an identical public furore, increased discussion of the scale and content of collaboration is a well-documented phenomenon (Conan and Rousso, 1994). Particularly during the series of trials for crimes against humanity, from the imprisonment of Klaus Barbie to the more recent incarceration of Maurice Papon, the French have been said to be obsessed with what they call the *années noires*. Although it is very difficult to measure what constitutes an 'obsession', it is fair to say that issues relating to Vichy have been prominent for many years, marking a range of areas within French society, including literary publishing, film, politics, journalism, historiography, high and popular culture.

Reviewing the present European scene, it would be correct to perceive a renewed concern in the history of the wartime period. This trend is as true for the lesser European states as it is for the old Great Powers. Even Switzerland has experienced major public questioning of the status of its neutrality between 1939 and 1945. The issue of the role played by its banking establishment in the handling of accounts held by victims of the Holocaust is another example of what could, if one liked the terminology, be called 'a return of memory'. Similarly, on the western fringe of the continent, in Great Britain, there have been some signs of a renewed interest in the history of Oswald Mosley and the British Union of Fascists. In middle-brow television dramas or what constitutes the British film industry, Blackshirts, appeasers and other fellow-travellers of the Right are now frequently represented on our screens.[3] Documentaries charting World Wars I and II are extant. It is thus credible to follow Judt and to agree that today, on the face of it, Europeans share a compelling interest in fascism and the Nazi period.

But, despite these illustrations, it is difficult to say that fascination with the war has gained pace in the post-1989 period. In fact, it is more persuasive to identify this process with the growth during the 1970s of interest in fascist memorabilia, fashion, art and culture.

Known in France as the '*mode rétro*', European cinema and literature saw a welter of products that re-examined the Third Reich. Works such as Visconti's *The Damned* (1968), Cavani's *The Night Porter* (1973) and Malle's *Lacombe Lucien* (1973), already presented the far right-wing in new ways. They spoke of fascism in stylish, elegant filmic languages that neither supported nor condemned history. Given this evidence, Judt may be exaggerating when he infers that everything changed with the fall of the Berlin wall. Reunification of Germany did focus minds back to the 1930s, but this was not new and cannot be firmly linked to the post-communist mood. Instead, it was probably part of a far longer trend in which intellectuals, politicians, journalists, novelists and others had, over several years, explored the meanings of Nazi history and its contemporary frisson.[4]

It is especially hard to demonstrate that the 'return of memory' (either before or after 1989) has contributed to the creation of a European cultural climate in which the rhetoric of nationalism has been re-legitimated. The fact that more Europeans were interested in issues related to the wartime past than ever before does not translate to a revival of the idea of the nation-state. Likewise it is tendentious to associate this shift in politics of memory with a decline in support of the EU. It would be more credible to claim merely that from the 1970s onwards there have been signs of a return to popular acceptance of intellectuals and ideological groups associated with the far right. For instance, returning to France, it is striking that the rise of the *Front National* has occurred precisely in the period which followed the original *mode rétro*'s exploration of Vichy. To the south and east of Europe, in Italy and Austria, there have been comparable re-awakenings of the far right, currently via the political activities of Haider and Fini. Thus, following the logic of Richard Golsan, we are perhaps watching 'Fascism's Return' (1998). However, it remains debatable whether there is a revival of nationalist sentiment beyond the extremes. In short, the mainstream of public opinion is probably unconcerned with either bolstering the nation or the birth of a federal super-state.

One can also underline that, 1989 notwithstanding, there has been a strong continuation of a discourse which reminds Europeans of the horrors of the past in ways which would hardly encourage nationalist breakthroughs. This interpretation is true for the British television documentaries and film dramas to which I have already referred, as well as for mass American products such as *Schindler's List* (1993) or more recently *Saving Private Ryan* (1998). Both of these offer representations which show why West European nationalism was so costly. These international films do not serve to revitalize a language of nationalism or associated militancy. On the contrary, the latest Spielberg production offers a harrowing – pseudo-documentary based – reminder of the violence of European state conflicts.

The geography of memory

A further area of enquiry prompted by Judt's work is the suggestion that experiences and memories of the war united Western Europe and distinguished it from Soviet Central and Eastern Europe. A geography, or cartography, of memory is implied. The argument that 1945 and 1989 witnessed a unified forgetting provides a coherent divide between the West European states and outsiders. The 'space' of Europe is thus potentially marked out by its willingness not to dwell on the past. Equally, Judt's discussion of post-1989 historical memory is again used as a marker of the West European area, albeit this time suggesting the return of memory and consequent fragmentation.

How plausible is this geographical picture of historical memory? Can one talk about 'West European' memory, and thereby also its counterpart 'Non-West European' memory? On one level of generality one can identify common attitudes towards the past which bring Europeans together. Typically, this would include the general cultural trend to learn from historical precedents (rather than to experiment based on ahistorical models). Equally, disregarding qualitative evidence, there *has* been a common interest in the meaning of the two World Wars. However, beyond the comparability at this level, one has to accept a far more pluralistic picture. First, there are varying attitudes towards the conflict which appear highly dependent on national experiences. These differences did not vanish in the 1950s. For example, returning to our introductory text, Bing's review of German history-writing and its national culture is a straightforward example which shows that national perspectives towards the past continued. The British historian is typical of his educational and social background, as well as characteristically *English* in his tone.

On the other hand, the Germans that he described were equally bound by their history and their interpretation of it. National communities established very different perspectives on what the past meant. Gaullist France, post-war Britain, West Germany, or any of the other nation-states took different views of what the war meant for them: not least because its most dramatic moments occurred at different times, for different reasons and through radically different historical episodes. For example, how many outside of France have recalled the Free French victory of Bir-Hakeim or celebrated the Second Tank Division which 'liberated' Paris? Similarly, very few in Germany will look at the Dunkirk escape in the way in which it is interpreted as an iconic moment in Britain. What does 'the bombing of Coventry' mean to an East European intellectual? Without listing further instances, we can see that memory still remained, and probably still remains, nationalized. If memory draws borders, it has long been critical in sustaining national differences, rather than establishing continental unity.

Looking beyond merely national divisions, it is also plain enough that the Jewish experience of 1939–45 was exceptionally traumatic and thus has produced an acute memory-reaction. Here, there is an international memory of the war which focuses on the Holocaust. From the Holocaust Museum in Washington DC to issues of Israeli national identity, Jewish memory of the war, with its global resonances, is distinct. Although taking place in Europe, the genocide of the Jews is usually seen as an event which goes beyond traditional boundaries altogether, and which carries universal lessons. It is both a European and a non-European legacy of Nazism which certainly does not serve to define a West European space. Furthermore, there are many other memories which work against common European attitudes. Briefly, we can see regional variations where memories of the war are different from the nation-states: Alsace-Lorraine, for instance, where occupation meant full incorporation into the Reich, with concomitant German military service. Likewise, in Northern Ireland, the Unionist community sees itself marching into battle for the defence of the realm, but elements within the wider Irish republican tradition, for instance the writer Francis Stuart were sympathetic to fascism.

Disregarding formal political borders there are urban and rural perspectives, male and female memories, left- and right-wing interpretations, and generational variations. There is a plurality of outlooks on contemporary history, which although bearing some similarities, tend also to establish difference. Even within national communities there is an almost perpetual diversity of historical memories. Consequently, there are sub-boundaries within national borders which divide citizens from each other. Despite the best efforts of unitary nation states to foster national memories – through staged commemorations, public holidays, or patriotic anthems which re-interpret victories and defeats – there is little evidence that these processes genuinely rally peoples to regard history through a *single* interpretation. There are inevitably variations and nuances of understanding based on gender, class, religion, region and politics. For short periods of time, state-led mass media events will encourage and persuade audiences to interpret historical events via a dominant ideology. But in the light of deep structures of difference for how long will this perspective retain its power?

If one had an imaginary map to replace the version set-up in Judt's analysis it would be a chaotic patchwork, no doubt with blurring but also distinctive groupings. Of course, taken in the round, the Central and Eastern European experiences would be kept apart from those of the West. Germany, for its part, was divided into two by ideology and still struggles to find solidarity. Westward, the Low Countries form another unit – but which itself would include different positions related to regionalism, religion, politics and gender. France and Britain are likewise separated from each other by the past. To the south, Italy remains strongly marked by fascist/anti-fascist debates in ways which appear unique. As for Spain and Portugal, their positions are exceptional due to the legacies of Salazar and Franco. In short, historical memories of conflict have tended to maintain divisions rather than to bridge them. Within the United Kingdom itself the Unionist community in Northern Ireland even today sets more store by the Protestant–Catholic divide than by the common Christianity of 'Europe' – a fact that surfaces in their particular kind of suspicion for EU institutions. For, just as historical memory in different locations and groupings is distinct, so are the borders they imply.

Historical memory is a slippery subject. Attractive and fascinating, it is not easy to use to explain major political developments, although it seems to partially shape them. I have examined Judt's use of memory, as part of his explanation for post-war European integration and as a defining feature of the continent. Whilst more questions have been asked than answered, impressionistically we can note that integration in Western Europe, as in the EDC debates in France, has confronted many divergent memories of the war. However, one can identify no clear periodization of memory of the kind Judt charts. The major patterns which he saw appear to be illusory. Instead of a continent united by its attitude towards the past we have found a divided set of reactions to World War II, dependent on numerous factors – national or regional culture, political, religious, class and gender differences. There are overlapping and interlocking memories, but still – perhaps in the nature of historical memory itself – little prospect of a single outlook on the history of Europe.

Notes

1 Judt's teasing style is reminiscent of his previous work on the French post-war intellectual circles of the Left Bank (1992). There he critiqued the cultural elite, in particular Sartre, for their philocommunist commitments. In the publication with which we are concerned, the idea of Europe and European integration appears to have offered a comparable set of reassuring certainties and arguments as Marx and subsequently Mao provided for elements of the left.

2 This is neither the time nor the place to discuss the relative merits of Rousso's work on French historical memory. Nevertheless, it is important to note that the argument which Judt's work supports, and extensively draws on, has been revised by, among others, Gordon (1995, 1998) and Lavabre (1987).

3 I have in mind productions such as the lavish Channel 4 series *Mosley* (1998), which was partially based on the biography of the fascist leader written by his son (republished 1998); as well as the adaptation of the novel *A Dance to the Music of Time* (1997) taken from the book-cycle written by Anthony Powell. In cinema, *Richard III* (1996) is an evocative portrayal of a fictional inter-war Britain in which the infamous Shakespearean hunchback is portrayed as a fascist villain.

4 Judt is aware of these developments, observing that works such as the 1970s historical treatment by Robert Paxton, *Vichy France* were influential taboo-breakers (p. 84). However, he remains unwilling to qualify the overarching contrast between periods which frames his argument.

References

Ayçoberry, P. (1981) *The Nazi Question: An Essay on the Interpretation of National Socialism (1922–1975)* (London: Routledge).

Azéma, Jean-Pierre and Bédarida, François (1994) 'L'Historisation de la résistance', *Esprit*, **198**(1), 19–35.

Bing, Harold F. (1951) 'The Study and Teaching of History in Post-War Germany', *History: The Journal of the Historical Association*, n.s. vol. XXXVI, no. 126/127, pp. 92–107.

Bosworth, R. J. B. (1992) *Explaining Auschwitz and Hiroshima: History-Writing and the Second World War* (London: Routledge).

Conan, E. and Rousso, Henry (1994) *Vichy: Un passé qui ne passe pas* (Paris: Fayard).

Frey, Hugo (1999) 'Re-building France: Gaullist Historiography, the Rise–Fall Myth and French Identity (1945–1958)', in S. Berger, M. Donovan and K. Passmore (eds), *Writing the Nation-State: Historians and National Identity in Europe Since 1800* (London: Routledge), pp. 205–16.

Golsan, R. J. (ed.) (1998) *Fascism's Return: Scandal, Revision and Ideology since 1980* (Lincoln, NA: University of Nebraska Press).

Gordon, Bertram (1995) 'The "Vichy Syndrome" Problem in History', *French Historical Studies*, **19**(2), 495–518.

Gordon, Bertram (1998) 'World War II France Half a Century After', in R. J. Golsan (ed.), *Fascism's Return: Scandal, Revision and Ideology since 1980* (Lincoln, NA: University of Nebraska Press), pp. 152–81.

Hartmann, Geoff (1986) *Bitburg in Moral and Political Perspective* (Bloomington, IN: Indiana University Press).

Judt, Tony (1992) *Past Imperfect: French Intellectuals, 1944–1956* (Berkeley, CA: University of California Press).

Judt, Tony (1997) *A Grand Illusion? An Essay on Europe* (London: Penguin).

Lavabre, Marie-Claire (1987) 'Du poids et du choix du passé. Lecture critique du Syndrome de Vichy', *Cahiers de l'IHTP*, **18**, 105–14.

Lacouture, Jean (1992) *De Gaulle*, vol. 2 (London: Harvill).

Maier, Charles S. (1988) *The Unmasterable Past: History, Holocaust and German National Identity* (Cambridge, MA: Harvard University Press).

Milward, Alan S. (1984) *The Reconstruction of Western Europe, 1945–51* (London: Methuen).

Milward, Alan S. (1992) *The European Rescue of the Nation-state* (London: Routledge).

Morris, Alan (1992) *Collaboration and Resistance Reviewed: Writers and the Mode Retro in Post-Gaullist France* (Oxford: Berg).

Mosley, Nicholas (1998) *Rules of the Game/Beyond the Pale: Memoirs of Sir Oswald Mosley and Family* (London: Pimlico).

Rioux, Jean-Pierre (1984) 'L'Opinion publique française et la CED: querelle partisane ou bataille de la mémoire', *Relations internationales*, vol. 37, pp. 37–53

Rioux, Jean-Pierre (1987) *The Fourth Republic, 1944–1958* (Cambridge: Cambridge University Press).

Rousso, H. (1990) *Le Syndrome de Vichy* (Paris: Seuil).

Conclusions

The discussions from which these chapters emerged began with the idea that the margins the integrated Europe has constructed in its own mind had unperceived effects on integration in Europe as a whole. After so many instances and explorations of the possibilities, it appears that marginality has indeed a number of specific effects which are too easily overlooked in the conventional, centre-oriented picture of 'Europe'.

Yet these are by no means one-way powers of the margins over integrated Europe as a whole. This book does not condone a view that the margins are determining what happens at the core, any more than it sanctions the converse view: that everything is decided from the centre. Shrewdly used, being marginal may give a lever in relation to others: British stand-offishness sometimes pays off; and politics in Iberia may give lessons to the whole of the EU. But in other instances, such as the Czech Republic, marginality redounds on the marginal: thus Czech optimists' sense that Western Europe owed them recognition courted put-downs such as Hans van den Broek's remark that 'it is not the European Union which wants to join the Czech Republic'. The developments on what Ann Kennard calls the boundary of Europe with 'the greatest multiplicity of marginal attributes' suggest another possibility. On the long-disputed Polish–German border, the principles of governance inherited from the two national traditions are themselves at odds, while some forces on the ground pull away from governmental management altogether. We may be looking at a permanent margin on the inside of integrated Europe.

At least as often as the margin's imposing on the whole, it is, however, the centre's anxieties about the margins which puts pressure on one, or the other, or both – as when the EU attempts to manoeuvre Mediterranean countries into trading relations and political understandings that their situation does not allow, with little satisfaction on either side. Or, again, there are those instances where what happens at the margins impacts on the centre without the margins knowingly intending that it should. The future effects of Central European countries' integration in the West will include, it seems, restructuring industry in Northern Europe and running-down the CAP. The latter is clearly not at all wished for by the marginal countries of Central Europe – though neither is it solely a product of the impact of that margin. The point about margins, it seems, is that they are hard to make fall into line: with effects both witting and unwitting, both expected and unexpected, upon integrated Europe as a whole.

Marginality is not an absolute category, but something which is attributed to areas and social actors (and interpreted, and contested) according to the power of various social orders. The studies in this volume indicate both the scope for manoeuvre in marginality and the difficulties. On the terrain of economic relations there are clear grounds to suspect that market processes tend to *produce* marginality, with a potential for harm on the whole. Yet strategies for reversing these tendencies are inherently disputable and often prove ineffective. Where we look at the more complex political and cultural levels we find that marginality can be played up or played down in different ways, with different effects, sometimes benign and sometimes disruptive. Margins are precisely that which it is 'hard to make fall into line', which is why we should at the very least keep the pulls from the margins firmly in view as we try to comprehend an ever-larger, if not an 'ever-closer' integrated Europe.

Index

Note: page numbers in italic indicate a table or figure.